T0253844

Gate Stack and Silicide
Issues in Silicon Processing

MATERIALS RESEARCH SOCIETY
SYMPOSIUM PROCEEDINGS VOLUME 611

Gate Stack and Silicide Issues in Silicon Processing

Symposium held April 25–27, 2000, San Francisco, California, U.S.A.

EDITORS:

L.A. Clevenger
International Business Machines
Hopewell Junction, New York, U.S.A.

S.A. Campbell
University of Minnesota
Minneapolis, Minnesota, U.S.A.

P.R. Besser
Advanced Micro Devices, Inc.
Austin, Texas, U.S.A.

S.B. Herner
Matrix Semiconductor
Santa Clara, California, U.S.A.

J. Kittl
Texas Instruments, Inc.
Dallas, Texas, U.S.A.

Materials Research Society
Warrendale, Pennsylvania

CAMBRIDGE UNIVERSITY PRESS
Cambridge, New York, Melbourne, Madrid, Cape Town,
Singapore, São Paulo, Delhi, Mexico City

Cambridge University Press
32 Avenue of the Americas, New York NY 10013-2473, USA

Published in the United States of America by Cambridge University Press, New York

www.cambridge.org
Information on this title: www.cambridge.org/9781107413160

Materials Research Society
506 Keystone Drive, Warrendale, PA 15086
http://www.mrs.org

First published 2001
First paperback edition 2013

Single article reprints from this publication are available through
University Microfilms Inc., 300 North Zeeb Road, Ann Arbor, MI 48106

CODEN: MRSPDH

ISBN 978-1-107-41316-0 Paperback

CONTENTS

HIGH-k MATERIALS

NOVEL GATE INSULATORS

NOVEL GATE STRUCTURES

ADVANCED GATE DIELECTRICS

INTEGRATION ISSUES IN THE FEOL

*Invited Paper

NOVEL SILICIDE PROCESSING

POSTER SESSION

SILICIDE FORMATION MECHANISMS

SHALLOW JUNCTIONS AND SILICIDES

*Invited Paper

EPITAXIAL SILICIDES

Author Index

Subject Index

· PREFACE

This volume contains papers presented at the three-day symposium "Gate Stack and Silicide Issues in Silicon Processing," held April 25–27 at the 2000 MRS Spring Meeting in San Francisco, California. This was the first MRS symposium dedicated solely to these issues. As we move into the 21st century, the feature size of microelectronic devices is approaching the deep submicron regime. At these dimensions, the process development and integration issues related to gate stack and silicide processing are key challenges. The gate leakage is rising sharply due to direct tunneling. Power and reliability concerns are both expected to limit the ultimate scaling of SiO_2 based insulators to about 1.5 nm. New gate insulators must not deleteriously affect the interface quality, thermal stability, charge trapping, or process integration. New metal gate materials and damascene gates are being seriously investigated, perhaps in conjunction with the application of a high permittivity gate insulator to provide sufficient device performance at ULSI dimensions. The silicidation process is also coming under very strong pressure. Narrow device widths and decreasing junction depths are making the formation of low leakage, low resistance silicide straps extremely difficult. Producing shallower junctions via ion implantation is inhibited by transient enhanced diffusion and low beam currents at low implantation energies. Gate stack and contact film effects, such as point defect injection, extended defect formation, and stress on ultrashallow junction formation must be considered. To be successful in creating the next generation of advanced Si technology, these issues related to gate stack and silicide processing must be addressed. This volume has collected many relevant papers in the areas of gate stack and silicide processing.

The financial support provided by Advanced Micro Devices, Applied Materials, and International Business Machines is acknowledged.

L.A. Clevenger
S.A. Campbell
P.R. Besser
S.B. Herner
J. Kittl

October 2000

MATERIALS RESEARCH SOCIETY SYMPOSIUM PROCEEDINGS

Volume 578— Multiscale Phenomena in Materials—Experiments and Modeling, I.M. Robertson, D.H. Lassila, R. Phillips, B. Devincre, 2000, ISBN: 1-55899-486-6

Volume 579— The Optical Properties of Materials, J.R. Chelikowsky, S.G. Louie, G. Martinez, E.L. Shirley, 2000, ISBN: 1-55899-487-4

Volume 580— Nucleation and Growth Processes in Materials, A. Gonis, P.E.A. Turchi, A.J. Ardell, 2000, ISBN: 1-55899-488-2

Volume 581— Nanophase and Nanocomposite Materials III, S. Komarneni, J.C. Parker, H. Hahn, 2000, ISBN: 1-55899-489-0

Volume 582— Molecular Electronics, S.T. Pantelides, M.A. Reed, J. Murday, A. Aviram, 2000, ISBN: 1-55899-490-4

Volume 583— Self-Organized Processes in Semiconductor Alloys, A. Mascarenhas, D. Follstaedt, T. Suzuki, B. Joyce, 2000, ISBN: 1-55899-491-2

Volume 584— Materials Issues and Modeling for Device Nanofabrication, L. Merhari, L.T. Wille, K.E. Gonsalves, M.F. Gyure, S. Matsui, L.J. Whitman, 2000, ISBN: 1-55899-492-0

Volume 585— Fundamental Mechanisms of Low-Energy-Beam-Modified Surface Growth and Processing, S. Moss, E.H. Chason, B.H. Cooper, T. Diaz de la Rubia, J.M.E. Harper, R. Murti, 2000, ISBN: 1-55899-493-9

Volume 586— Interfacial Engineering for Optimized Properties II, C.B. Carter, E.L. Hall, S.R. Nutt, C.L. Briant, 2000, ISBN: 1-55899-494-7

Volume 587— Substrate Engineering—Paving the Way to Epitaxy, D. Norton, D. Schlom, N. Newman, D. Matthiesen, 2000, ISBN: 1-55899-495-5

Volume 588— Optical Microstructural Characterization of Semiconductors, M.S. Unlu, J. Piqueras, N.M. Kalkhoran, T. Sekiguchi, 2000, ISBN: 1-55899-496-3

Volume 589— Advances in Materials Problem Solving with the Electron Microscope, J. Bentley, U. Dahmen, C. Allen, I. Petrov, 2000, ISBN: 1-55899-497-1

Volume 590— Applications of Synchrotron Radiation Techniques to Materials Science V, S.R. Stock, S.M. Mini, D.L. Perry, 2000, ISBN: 1-55899-498-X

Volume 591— Nondestructive Methods for Materials Characterization, G.Y. Baaklini, N. Meyendorf, T.E. Matikas, R.S. Gilmore, 2000, ISBN: 1-55899-499-8

Volume 592— Structure and Electronic Properties of Ultrathin Dielectric Films on Silicon and Related Structures, D.A. Buchanan, A.H. Edwards, H.J. von Bardeleben, T. Hattori, 2000, ISBN: 1-55899-500-5

Volume 593— Amorphous and Nanostructured Carbon, J.P. Sullivan, J. Robertson, O. Zhou, T.B. Allen, B.F. Coll, 2000, ISBN: 1-55899-501-3

Volume 594— Thin Films—Stresses and Mechanical Properties VIII, R. Vinci, O. Kraft, N. Moody, P. Besser, E. Shaffer II, 2000, ISBN: 1-55899-502-1

Volume 595— GaN and Related Alloys—1999, T.H. Myers, R.M. Feenstra, M.S. Shur, H. Amano, 2000, ISBN: 1-55899-503-X

Volume 596— Ferroelectric Thin Films VIII, R.W. Schwartz, P.C. McIntyre, Y. Miyasaka, S.R. Summerfelt, D. Wouters, 2000, ISBN: 1-55899-504-8

Volume 597— Thin Films for Optical Waveguide Devices and Materials for Optical Limiting, K. Nashimoto, R. Pachter, B.W. Wessels, J. Shmulovich, A.K-Y. Jen, K. Lewis, R. Sutherland, J.W. Perry, 2000, ISBN: 1-55899-505-6

Volume 598— Electrical, Optical, and Magnetic Properties of Organic Solid-State Materials V, S. Ermer, J.R. Reynolds, J.W. Perry, A.K-Y. Jen, Z. Bao, 2000, ISBN: 1-55899-506-4

Volume 599— Mineralization in Natural and Synthetic Biomaterials, P. Li, P. Calvert, T. Kokubo, R.J. Levy, C. Scheid, 2000, ISBN: 1-55899-507-2

Volume 600— Electroactive Polymers (EAP), Q.M. Zhang, T. Furukawa, Y. Bar-Cohen, J. Scheinbeim, 2000, ISBN: 1-55899-508-0

Volume 601— Superplasticity—Current Status and Future Potential, P.B. Berbon, M.Z. Berbon, T. Sakuma, T.G. Langdon, 2000, ISBN: 1-55899-509-9

Volume 602— Magnetoresistive Oxides and Related Materials, M. Rzchowski, M. Kawasaki, A.J. Millis, M. Rajeswari, S. von Molnár, 2000, ISBN: 1-55899-510-2

Volume 603— Materials Issues for Tunable RF and Microwave Devices, Q. Jia, F.A. Miranda, D.E. Oates, X. Xi, 2000, ISBN: 1-55899-511-0

Volume 604— Materials for Smart Systems III, M. Wun-Fogle, K. Uchino, Y. Ito, R. Gotthardt, 2000, ISBN: 1-55899-512-9

MATERIALS RESEARCH SOCIETY SYMPOSIUM PROCEEDINGS

Prior Materials Research Society Symposium Proceedings available by contacting Materials Research Society

High-k Materials

Mat. Res. Soc. Symp. Vol. 611 © 2000 Materials Research Society

The Structure of Plasma-Deposited and Annealed Pseudo-Binary ZrO_2-SiO_2 Alloys

G. Rayner Jr., R. Therrien, and G. Lucovsky

Departments of Physics, Electrical and Computer Engineering, and Materials Science and Engineering
North Carolina State University, Raleigh, NC 27695-8202, USA

ABSTRACT

The internal structures of various $(ZrO_2)_x(SiO_2)_{1-x}$ alloys ($x \leq 0.5$) were investigated. A remote plasma enhanced-metal organic chemical vapor deposition (RPE-MOCVD) process was used to deposit films with varying alloy composition on Si(100) substrates. This study indicates that for the *glassy* silicate phase, g-$ZrSiO_4$, a glass transition temperature, T_g, exists between 800°C and 900°C at which phase separation into the end-member components, SiO_2 and ZrO_2, occurs.

INTRODUCTION

Pseudo-binary $(ZrO_2)_x(SiO_2)_{1-x}$ alloys are currently being considered as a high-k dielectric for microelectronic applications [1,2]. Due to post-deposition processing, one important issue is thermal stability of the internal structure. This paper identifies some limitations of these alloys for high-k dielectric applications. These limitations are associated with the thermal stability of the stoichiometric silicate phase, $ZrSiO_4$.

EXPERIMENTAL

Films were deposited by a RPE-MOCVD process described elsewhere [1] onto Si(100) substrates. Film composition was determined by Rutherford backscattering spectrometry (RBS). Alloy internal structure, as-deposited and following subsequent annealing, was characterized by X-ray diffraction (XRD) and infrared (IR) absorption spectroscopy. Post deposition 60s anneals at 600°C, 700°C, 800°C, and 900°C were performed *ex-situ* on each sample in an inert argon ambient using an AG Minipulse 310 RTA unit. XRD measurements were made using a Bruker x-ray diffractometer with a beryllium area detector centered at the following 2θ positions: 25° and 50°. IR absorption measurements in the mid-IR regime (4000-400cm^{-1}) were carried out on a Nicolet Magna-FTIR 750 spectrometer. Due to the relatively slow deposition rate (~ 4 Å/min), films were deposited for eight hours to ensure that the thickness was sufficient for good IR absorption sensitivity.

RESULTS AND DISCUSSION

A portion of the equilibrium compositional phase diagram [3] for the pseudo-binary ZrO_2-SiO_2 system is presented in Fig. 1. This figure illustrates only the chemical phase of each component and not any particular structural phase. The phase diagram, for temperatures below ~1600°C, can be divided into two regions based on chemical composition: 1) a SiO_2-rich region $(0 \leq x < 0.5)$ from SiO_2 to the silicate phase, $ZrSiO_4$, and 2) a ZrO_2-rich region $(0.5 < x \leq 1)$ from the silicate phase to the end-member, ZrO_2. Above ~1600°C, $ZrSiO_4$ phase separates into the end-member components, SiO_2 and ZrO_2 $(ZrSiO_4 \rightarrow ZrO_2 + SiO_2)$. It must be emphasized that this transition temperature corresponds to the crystalline silicate phase, c-$ZrSiO_4$. Included in this diagram is the glass transition temperature, T_g, which corresponds to the temperature at which the *glassy* silicate phase, g-$ZrSiO_4$, softens. This aspect of the phase diagram will be discussed later.

Rutherford Backscattering Spectrometry

Film composition as a function of depth was determined by Rutherford backscattering spectrometry. This study investigated three ZrO_2-SiO_2 alloys. RBS results indicate that, within experimental uncertainty, both species are fully oxidized (i.e., %O = 66.7%, %Si + %Zr = 33.3%) having only Si-O and Zr-O bonds. RBS depth profiling reveals that ZrO_2-composition varies as a function of depth in each film. The data shows that ZrO_2-compositon increases as a function of film depth having the highest composition near the Si-interface. With increasing ZrO_2 incorporation variations in composition as a function of depth become more significant. For each $(ZrO_2)_x(SiO_2)_{1-x}$ alloy, Table I lists the average value of x and the overall change in x, Δx. The overall change in composition, Δx, is the measured value of x near the Si-interface minus the value near the film surface. Also included in Table 1 is the compositional uncertainty for each species. According to RBS, films are approximately 2000 Å thick.

X-Ray Diffraction

The XRD results of Zr10 are presented in Fig. 2a. As-deposited, a weak, broad feature is observed at a 2θ value of approximately 23° indicating the film is amorphous. No changes in structure are observed after annealing. Similarly, XRD results of the alloy with intermediate ZrO_2-composition, Zr20, in Fig. 2b indicate a stable amorphous structure. The XRD data of as-deposited Zr50 in Fig. 2c has a relatively strong, broad feature at approximately 30° indicative of an amorphous phase. No changes in structure are observed when annealed up to 800°C; however, crystallization occurred at 900°C, and the structural phase was identified as tetragonal-ZrO_2.

Infrared Absorption Spectroscopy

The IR absorption spectra of the two end-members in Fig. 1, SiO_2 and ZrO_2, can be used to establish a basis for understanding the IR response of the ZrO_2-SiO_2 alloy system. IR spectra of the two end-members, both as-deposited and following a 900°C anneal, are presented in Fig. 3. The local bonding in SiO_2 is made up entirely of covalent Si-O-Si bonds. In general, covalent-oxide glasses form continuous random network (CRN) structures that are stable when heated to relatively high temperatures. The IR spectra of the a-SiO_2 film in Fig. 3 has three characteristic features: i) a stretching-mode vibration at ~ 1070 cm^{-1}, ii) a bending-mode vibration at ~ 810 cm^{-1}, and iii) a out-of-plane rocking-mode vibration at ~ 455 cm^{-1}. Each Si-atom in the network has four equivalent oxygen nearest neighbors which results in the high wavenumber shoulder at ~ 1200 cm^{-1}. After annealing at 900°C there is no significant change in the IR response indicating the internal structure is thermodynamically stable.

Local bonding in ZrO$_2$ is characterized by ionic Zr-O-Zr bonds. In contrast to covalent-oxide glasses, ionic-oxide glasses form random close packed (RCP) structures which are not thermodynamically stable (i.e., they readily crystallizes when relatively small amounts of energy in the form of heat is added). The as-deposited film in Fig. 3 has spectral features associated with a polycrystalline monoclinic phase of ZrO$_2$, as determined by XRD. Note that these features, which are due to stretching-mode vibrations of ionic Zr-O bonds, are observed in the region from ~ 400-800 cm^{-1}. The IR response remains unchanged following a 900°C anneal indicating no changes in internal structure.

The IR absorption spectra of Zr10 in Fig. 4a has features at ~ 1067 cm^{-1}, ~ 810 cm^{-1}, and ~ 450 cm^{-1} that are indicative of an a-SiO$_2$ covalent CRN structure previously discussed. The incorporation of ZrO$_2$ into the a-SiO$_2$ matrix results in the formation of Si-O-Zr bonds. Since the bonding between zirconium and oxygen, according to Pauling electronegativity, is mainly ionic, the Zr-atoms act as a network modifier and not as part of the network structure. Therefore, the shoulder observed at ~ 950 cm^{-1} is assigned to a stretching-mode vibration of the non-bridging oxygen atoms, Si-O^{-1}, associated with Si-O-Zr bonds. The IR response after successive anneals indicates that the internal structure is relatively stable.

IR results of Zr20 in Fig. 4b are similar to those obtained for Zr10. However, as a result of increased ZrO$_2$ composition, changes in the internal structure related to ZrO$_2$ incorporation into the a-SiO$_2$ matrix now becomes more apparent. Specifically, a broadening of the shoulder at ~ 950 cm^{-1} and a broad band of modes extending from ~ 400-800 cm^{-1} are observed due to the increase of Si-O-Zr bonding. From the previous discussion on the IR response of c-ZrO$_2$ in Fig. 3, the band of modes from ~ 400-800 cm^{-1} are due to stretching-mode vibrations related to Zr-O bonds. The internal structure is relatively stable up to 800°C, but at 900°C modes assigned to Zr-O bonds show some evidence of structural change indicated by a weak shoulder at ~ 600 cm^{-1} in the IR response.

The IR absorption spectra of the alloy having the highest ZrO$_2$-composition, Zr50, is shown in figure 4c. For annealing temperatures up to 800°C no evidence of a bending-mode vibration at ~ 810cm^{-1} is observed indicating the absence of an a-SiO$_2$ CRN structure. Since no Zr-O stretching-modes are observed above ~ 800 cm^{-1} in Fig. 3, features from approximately 800-1200 cm^{-1} must be due to vibrational modes related to tetrahedral SiO$_4^{-4}$ molecular-ions. Features from approximately 400-800 cm^{-1} contain a superposition of modes related to SiO$_4^{-4}$ molecular-ions, and ionic Zr-O bonds. Therefore, the internal structure of the as-dep. film is mostly the *glassy* stoichiometric phase, g-ZrSiO$_4$, where each O-atom forms only Si-O-Zr bonds. After annealing up to 800°C, the internal structure is relatively stable. However, at 900°C the internal structure changes rather dramatically. The presence of a bending mode vibration at ~ 810 cm^{-1} indicates that an a-SiO$_2$ network is now present. Also, the large decrease in the intensity of modes ranging from ~ 800-1000 cm^{-1} indicates that a large number of Si-O-Si and Zr-O-Zr bonds were formed (i.e., 2·Si-O-Zr → Si-O-Si + Zr-O-Zr). Using XRD results in Fig. 3c, the features at ~ 450 cm^{-1}, and ~ 600 cm^{-1} are now dominated by tetragonal phase c-ZrO$_2$. The reduced number of sharp features, relative to c-ZrO$_2$ monoclinic-phase in Fig. 4, is consistent with the increased symmetry of the tetragonal-phase.

The transition from Si-O-Zr bonding to Si-O-Si and Zr-O-Zr bonding at 900°C in Zr50 indicates chemical separation of the g-ZrSiO$_4$ phase. The shoulder at 600cm^{-1} in Fig. 4c related to the chemical phase, ZrO$_2$, is also weakly observed in Zr20 following a 900°C anneal indicating that regions containing only Zr-O-Zr bonds are present in this alloy as well. Results indicate that the glass transition temperature, T$_g$, for the g-ZrSiO$_4$ phase is somewhere between 800°C and 900°C. At temperature, T$_g$, Si-O bonds begin to break allowing the g-ZrSiO$_4$ phase to separate into the end-member components, SiO$_2$ and ZrO$_2$. In the crystalline phase, this separation does not occur until ~1600°C.

CONCLUSIONS

The experimental results indicate that the as-deposited films are chemically ordered having only Si-O-Si and Si-O-Zr bonds consistent with the equilibrium phase diagram. Both IR and XRD indicate that the internal structure is relatively stable up to 800°C. However, at 900°C results indicate that g-ZrSiO$_4$ chemically phase separates forming regions containing only Si-O-Si and Zr-O-Zr bonds, with Si-O-Zr bonds occurring only at internal boundaries. These results indicate limitations of these alloys for high-k dielectric applications due to: 1.) chemical phase separation resulting in lower effective dielectric constants, and 2.) crystallization of regions containing ZrO$_2$ which can lead to high leakage currents. Further investigation is needed to determine how these results are related to film thickness.

ACKNOWLEDGMENTS

The authors would like to thank Jon-Paul Maria at NCSU for the use of the x-ray diffractometer. Funding for this research was provided by ONR, NSF, and SEMATECH/SRC Front End Processes Center.

REFERENCES

1. D. M. Wolfe, K. Flock, R. Therrien, R. Johnson, B. Rayner, L. Günther, N. Brown, B. Claflin, and G. Lucovsky, Mat. Res. Soc. Symp. Proc. **567**, 343 (1999).

2. G. D. Wilk, R. M. Wallace, and J. M. Anthony, J. Appl. Phys. **87**, 484 (2000).

3. E. M. Levin, C. R. Robbins, and H. F. McMurdie, Phase Diagrams for Ceramists, *Amer. Ceram. Soc.*, **v.2**, (1969)

Table I. RBS results indicate that ZrO_2-composition increases as a function of depth; therefore, only the average value of x is reported here. The overall change in composition, Δx, is the measured value of x near the Si-interface minus the value near the film surface.

Fig. 1 A portion of the equilibrium compositional phase diagram for the $(ZrO_2)_x(SiO_2)_{1-x}$ alloy system. The black dots represent the samples discussed in this paper.

Fig. 2a The XRD results of Zr10 as-deposited and after annealing. The weak, broad feature at ~ 23° indicates an amorphous structure.

Fig. 2b The XRD results of Zr20 as-deposited and after annealing.

Fig. 2c The XRD results of Zr50 as-deposited and after annealing. The relatively strong, broad feature at ~ 30° indicates an amorphous structure. However, at 900°C crystallization occurred and the structural phase was identified as tetragonal ZrO_2.

Fig. 3 IR absorption spectra of the two end-members components, SiO_2 and ZrO_2, as-deposited and after a 900°C anneal.

Fig. 4a IR absorption spectra of Zr10 as-deposited and after annealing. The shoulder observed at ~ 950 cm^{-1} is assigned to a stretching-mode vibration of the non-bridging oxygen atoms, Si-O^{-1}, associated with Si-O-Zr bonds.

Fig. 4b IR absorption spectra of Zr20 as-deposited and after annealing. At 900°C modes assigned to Zr-O bonds show evidence of structural change indicated by a weak shoulder at ~ 600 cm^{-1}.

Fig. 4c IR absorption spectra of Zr50 as-deposited and after annealing. At 900°C the presence of a bending mode vibration at ~ 810 cm^{-1} indicates an a-SiO_2 CRN structure.

RBS Results

Sample	x (Avg. ±1.5%)	Δx
Zr10	0.10	0.033
Zr20	0.229	0.069
Zr50	0.50	0.102

* Experimental Uncertainty: %O ±6%, %Si ±2%, %Zr ±0.5%

Table I

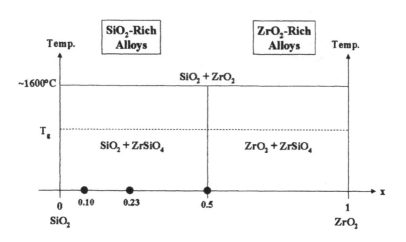

Fig. 1

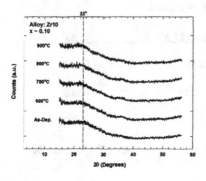

Fig. 2a

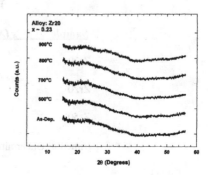

Fig. 2b

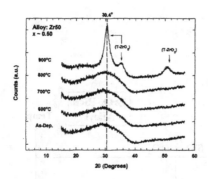

Fig. 2c

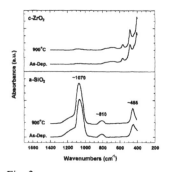

Fig. 3

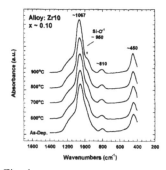

Fig. 4a

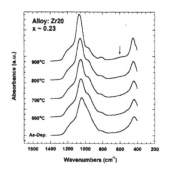

Fig. 4b

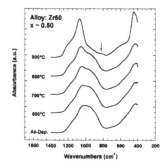

Fig. 4c

Mat. Res. Soc. Symp. Vol. 611 © 2000 Materials Research Society

Physical and Electrical Properties of Yttrium Silicate Thin Films

James J. Chambers and Gregory N. Parsons
Department of Chemical Engineering, North Carolina State University, Raleigh, NC 27695

ABSTRACT

This article reports on the physical and electrical properties of yttrium silicate, which is a possible high-k replacement for the SiO_2 gate dielectric in CMOS devices. The yttrium silicate (Y-O-Si) films are formed by sputtering yttrium onto clean silicon, annealing in vacuum to form yttrium silicide and then oxidizing in N_2O to form the silicate. Shifts in the Y 3d, Si 2p and O 1s photoelectron spectra with respect to Y_2O_3 and SiO_2 indicate that the films are fully oxidized yttrium silicate. FTIR results that reveal a Si-O stretching mode at 950 cm^{-1} and Y-O stretching modes in the far-IR are consistent with XPS. XPS and FTIR results are in accordance with the donation of electron density from the yttrium to the Si-O bond in the silicate. The yttrium silicate films contain a fixed charge density of $\sim 9 \times 10^{10}$ cm^{-2} negative charges as calculated from measured C-V behavior. The properties of ultra-thin yttrium silicate films with an equivalent silicon dioxide thickness (electrical) of ~ 1.0 nm will be discussed elsewhere.

INTRODUCTION

The continuous scaling of complementary-metal-oxide-semiconductor (CMOS) devices toward smaller dimensions will require replacing the gate SiO_2 with an alternate high dielectric constant (high-k) material. Metal silicates are appealing materials, since they are expected to posses a greater dielectric constant than SiO_2 and crystallize at higher annealing temperatures than metal oxides. Yttrium silicate possesses desirable thermodynamic, dielectric and structural properties that make it attractive as a high k candidate. The Y-O bond is quite strong, since the free energy of formation[1] (at 25 °C, per O-atom) is -2.40×10^{-22} kcal compared to -1.70×10^{-22} kcal for SiO_2. Previous researchers demonstrated MOS capacitors with an $Al/Y_{2.45}Si_{0.55}O_5/Si$ structure and obtained a dielectric constant for the silicate layer of ~ 12,[2] which should be suitably large to obtain EOT < 1.0 nm with low tunneling. Figure 1a depicts the mineral keiviite ($Y_2O_3 \cdot 2SiO_2$), which consists of a $Si_2O_7^{6-}$ structural unit with yttrium connecting the two corner sharing SiO_4 tetrahedra.[3] Figure 1b shows the yttrium orthosilicate ($Y_2O_3 \cdot SiO_2$) structure where four of the oxygen atoms are bound in a silicon tetrahedron with each corner joining two yttrium octahedron and the fifth oxygen atoms, which are not involved with the silicon tetrahedron, are shared between four yttrium octahedron in a rod-like chain.[4] $Y_2O_3 \cdot 2SiO_2$ and $Y_2O_3 \cdot SiO_2$ exhibit low lattice mismatch with silicon, since the lattice constants $a(Y_2O_3 \cdot 2SiO_2) = 0.554$ nm and $a(Y_2O_3 \cdot SiO_2) = 1.041$ nm are closely match with a (Si) = 0.543 and a(Si)x2 = 1.086 nm, respectively. Also, yttrium silicate films exhibit low oxygen permeability and should provide excellent passivation.[5] These advantageous properties make yttrium silicate worthy of study as a dielectric for CMOS technology.

We formed yttrium silicate films on silicon by oxidation of yttrium silicide, where the silicide films are grown by vacuum annealing of yttrium films on clean silicon. A similar approach has been utilized elsewhere to investigate the oxidation properties of silicides, where hafnium silicide was oxidized to from hafnium silicate.[6] Oxidation of certain metal silicides can result in SiO_2 growth over the silicide, as in the oxidation of nickel silicide and cobalt silicide.[7]

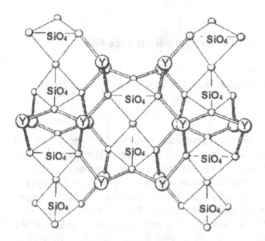
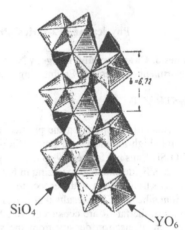

Figure 1a. Schematic depicting crystal
structure of $Y_2O_3 \cdot 2SiO_2$. (from Ref. 3)

Figure 1b. Schematic depicting crystal
structure of $Y_2O_3 \cdot SiO_2$. (from Ref. 4)

It has been postulated that oxidation of a metal silicide results in silicate formation if the metal-oxygen bond posses a more negative free energy of formation than the silicon-oxygen bond.[8] Since the free energy of formation for the Y-O bond is greater than that of the Si-O bond, oxidation of yttrium silicide is expected to result in silicate formation.

This article reports on the physical and electrical properties of yttrium silicate as compared to the properties of SiO_2 and Y_2O_3. X-ray photoelectron spectroscopy (XPS) is used to compare the oxidation state of yttrium, silicon and oxygen in yttrium silicate, SiO_2 and Y_2O_3. Fourier transform infrared spectroscopy (FTIR) is used to identify the Si-O and Y-O vibrational modes in yttrium silicate, SiO_2 and Y_2O_3. Yttrium silicate film topology is investigated using atomic force microscopy (AFM). The electrical quality of the yttrium silicate is tested using capacitance versus voltage (C-V) measurements.

EXPERIMENTAL

Substrates were p-type Si(100) with resistivity 0.1-0.3 Ω cm. Samples were prepared by dipping for 5 minutes in JTB 100 (a tetramethylammonium hydroxide based alkaline solution with a carboxylate buffer), rinsing in deionized water, etching in buffered hydrogen fluoride for 30 seconds with no final rinse and loading immediately into vacuum.

Yttrium silicate films were formed by sputtering yttrium onto clean silicon, vacuum annealing and then oxidizing in N_2O. The exact composition of the films reported here is unknown, and we denote our films as yttrium silicate or simply as Y-O-Si. Sputtering was performed at a substrate temperature of 25 °C in 4.3 mTorr Ar, a radio frequency (rf) power of 420 W at 13.56 MHz and an yttrium target dc bias of –1000 V in a system with an ultimate vacuum of 5×10^{-8} Torr. The yttrium films on silicon were annealed *in situ* in vacuum (< 1×10^{-6} Torr) at 600 °C to form yttrium silicide. The yttrium silicide was thermally oxidized *ex situ* in a standard tube furnace in 1 atm N_2O at 900 °C. SiO_2 films were thermally grown (~100 nm) by wet oxidation. Y_2O_3 films were formed by sputtering yttrium onto ~50 nm SiO_2 and then oxidizing at 900 °C in 1 atm N_2O.

Yttrium silicate films for XPS, FTIR and AFM analysis were formed by sputtering 240 nm of yttrium onto cleaned silicon, annealing in vacuum for 30 minutes and oxidizing in N_2O for 20 minutes. XPS was performed *ex situ* using a Riber LAS3000 utilizing a single-pass cylindrical mirror analyzer with an input lens and non-monochromatic Mg Kα ($h\nu = 1253.6$ eV). Spectra were collected using 0.1 eV step sizes and a resolution of ~1.0 eV. The binding energies of the core levels were corrected by setting the adventitious carbon 1s peak at 285 eV. FTIR spectra were collected using a Nicolet Magna 750 configured for both mid- and far-IR data acquisition. FTIR spectra were collected in a transmission mode, normalized to the silicon substrate and then converted to an absorbance scale. The thin SiO_2 buffer that the Y_2O_3 films were formed on was subtracted from the Y_2O_3 FTIR spectra. High resistivity (15 Ω-cm) silicon wafers were used for FTIR analysis to minimize free carrier absorption in the substrate. These substrates were frontside polished with an "orange peel" roughened backside to reduce background interference fringes.

Yttrium silicate films for electrical testing were formed by sputtering 2.5 nm of yttrium onto clean silicon, annealing in vacuum for 5 minutes and then oxidizing at 900 °C in N_2O for 15 seconds. Frontside electrical contact (no back-side metal employed) was made to the yttrium silicate films using a Four Dimensions CV map 92-A mercury probe. Capacitance measurements were taken at 1 MHz using a HP 4284a LCR meter.

RESULTS AND DISCUSSION

Chemical bonding in yttrium silicate, SiO_2 and Y_2O_3 was investigated using XPS. Figure 2 presents the Y 3d photoelectron spectra for yttrium silicate and Y_2O_3. The Y 3d peak is a doublet resulting from the Y $3d_{3/2}$ and the Y $3d_{5/2}$ (the more intense component where peak position is referenced) components of the spin-orbit splitting. The Y 3d peak for Y_2O_3 is measured at 156.6 eV binding energy (BE) compared to 156.7 eV for a reference[9] Y_2O_3. The Y 3d peak for yttrium silicate is measured at 157.6 eV.

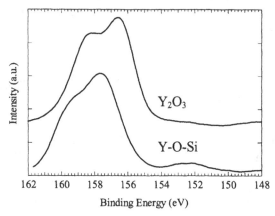

Figure 2. Y 3d portion of the XPS spectra for yttrium silicate and Y_2O_3.

The Y 3d spectra for yttrium silicate and Y_2O_3 in Figure 2 indicate that the yttrium in both films is fully oxidized, since a metallic Y 3d peak[10] at 156.0 eV is not observed. Based on the relative electronegativities of yttrium, silicon and oxygen (1.2, 1.8, 3.5 on the Pauling scale), the Y 3d BE for yttrium silicide is expected to lie near yttrium metal. There is no evidence for Y-Si bonds in the yttrium silicate Y 3d spectrum. Since yttrium and silicon are only bonded to oxygen in the yttrium silicate crystal structures exhibited in Figures 1a and 1b, the absence of Y-Si bonding in yttrium silicate films is expected. The Y 3d peak for yttrium silicate is shifted to higher BE than the Y 3d peak for Y_2O_3. This is likely due to increased donation of electron density to the Si-O bond in the silicate compared to the electron density donation from yttrium to oxygen in Y_2O_3, consistent with the relative electronegativities of yttrium, silicon and oxygen.

Figure 3 shows the Si 2p portion of the XPS signal for yttrium silicate and SiO_2. The Si 2p photoelectron peak position for SiO_2 is measured at 103.3 eV and reported for a reference[10] SiO_2 at 103.3 eV. The Si 2p photoelectron peak position for yttrium silicate is measured at 101.0 eV. The yttrium silicate Si 2p peak is shifted to lower BE than that for SiO_2 consistent with yttrium donating electron density to the Si-O bond. The observation that the Y 3d and Si 2p peak positions for the yttrium silicate are different than either Y 3d in Y_2O_3 or Si 2p in SiO_2 indicates the presence of a uniform material with the absence of phase separated Y_2O_3 or SiO_2.

Figure 4 presents the O 1s spectra for yttrium silicate, Y_2O_3 and SiO_2. The peak positions for the yttrium silicate, Y_2O_3 and SiO_2 are 530.5, 529.25 and 533.0 eV, respectively. The shoulder observed on the Y_2O_3 spectrum at 531.5 eV is assigned to O-H bonding in the film likely acquired during post-deposition exposure to ambient water.[9] The O 1s peak for yttrium silicate is measured to fall between Y_2O_3 and SiO_2, as expected considering the electronegativity differences between yttrium and silicon. Again, the absence of an O 1s peak for SiO_2 or Y_2O_3 in the yttrium silicate spectrum indicates the absence of any phase segregation in the silicate.

Analysis of the bonding environments in yttrium silicate, SiO_2 and Y_2O_3 were investigated using FTIR. Figure 5a presents the mid-IR region of the FTIR spectrum for yttrium silicate and Y_2O_3. The SiO_2 FTIR spectrum presented in Figure 5a exhibits three characteristic vibrational modes of SiO_2. The peaks located at 1077, 810 and 400 cm^{-1} are the Si-O-Si asymmetric stretching, bending and rocking modes, respectively. These peaks are typical for a stoichiometric and relaxed amorphous SiO_2 network. The dominant peak in the yttrium silicate FTIR spectrum in Figure 5a is at 950 cm^{-1}. Although the peak at 950 cm^{-1} in the yttrium silicate

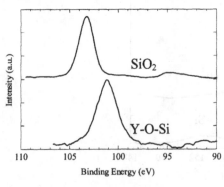

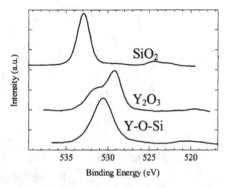

Figure 3. Si 2p portion of the XPS spectra for yttrium silicate and SiO_2.

Figure 4. O 1s portion of the XPS spectrum for yttrium silicate, Y_2O_3 and SiO_2.

FTIR spectrum is broader and shifted to lower frequency, it is similar in shape to the asymmetric stretch measured at 1077 cm^{-1} and is assigned to Si-O stretching in the yttrium silicate. The frequency shift and broadening of the Si-O stretch in the yttrium silicate is consistent with a weakening of the Si-O bond resulting from the combined effects of transfer of electron density from the yttrium to the Si-O bond and a wider distribution of Si-O bond angles available in the silicate. Figure 5b presents the far-IR region of the FTIR spectrum for yttrium silicate and Y_2O_3. The Y_2O_3 spectrum in Figure 5b exhibits vibrational modes at 374, 335 and 303 cm^{-1} characteristic of Y-O stretching in Y_2O_3.[11] The yttrium silicate spectrum in Figure 5b displays two broad peaks assigned to Y-O stretching in the silicate.

Figure 6 presents an AFM image of an yttrium silicate film with a root mean square surface roughness of 8.0 nm. It has been reported that vacuum annealing of yttrium films on silicon forms an epitaxial silicide with grain sizes on the order of 70 nm.[12] It is possible that the roughness inherent in the yttrium silicide remains intact during oxidation to yttrium silicate.

Capacitance-voltage testing of yttrium silicate films was performed to ascertain their electrical quality and the results are depicted in Figure 7. The equivalent oxide thickness (EOT)

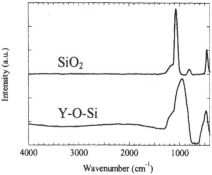

Figure 5a. Mid-IR region of the FTIR spectrum for yttrium silicate and SiO_2.

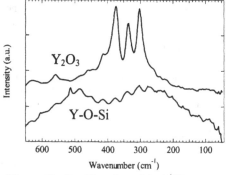

Figure 5b. Far-IR region of the FTIR spectrum for yttrium silicate and Y_2O_3.

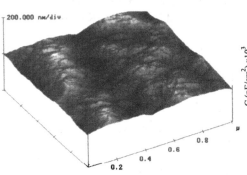

Figure 6. AFM image of yttrium silicate with RMS surface roughness of 8.0 nm.

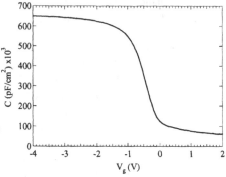

Figure 7. C-V curve for yttrium silicate with EOT = 5.0 nm.

for this film was calculated to be ~5.0 nm. The ideal flat band voltage (V_{fb}) for a mercury-oxide-semiconductor (N_a = 1×10^{17} cm^{-3}) capacitor is -0.55 V. V_{fb} for the C-V curve displayed in Figure 7 is measured at –0.51 V, which corresponds to a fixed charge density of ~9×10^{10} cm^{-2} and suggests the presence of negative fixed charges in the silicate.

CONCLUSIONS

We have described the physical and electrical properties of yttrium silicate films formed on silicon by oxidation of yttrium silicide. The shift of the Y 3d photoelectron peak to higher BE compared Y_2O_3, the shift of the Si 2p peak to lower BE compared to SiO_2 and the intermediate location of the O 1s peak compared to Y_2O_3 and SiO_2 indicate the formation of yttrium silicate. There is no evidence in the XPS spectra for the presence of SiO_2, Y_2O_3 or yttrium silicide. XPS spectra are consistent with the detection of yttrium silicate in the FTIR spectra where a Si-O stretching peak is measured at 950 cm^{-1} and Y-O stretching peaks are measured in the far-IR. These results are consistent with the donation of electron density from yttrium to the Si-O bond in the silicate. C-V measurements reveal that the yttrium silicate films contain ~9×10^{10} cm^{-2} fixed negative charge density. Based on these results, we have investigated ultra-thin (EOT $\cong$ 1.0 nm) yttrium silicate films and will report the results elsewhere.

ACKNOWLEDGMENTS

We acknowledge funding from the SRC/SEMATECH Center for Front End Processes at North Carolina State University. We thank Bruce Rayner and Dr. Gerald Lucovsky for their assistance in obtaining the far-IR spectra.

REFERENCES

1 I. Barin, *Thermochemical Data of Pure Substances* (VCH Verlagsgesellschaft, Weinheim, Germany, 1989).

2 M. Gurvitch, L. Manchanda, and J. M. Gibson, Appl. Phys. Lett. **51**, 919-921 (1987).

3 N. G. Batalieva and Y. A. Pyatenko, Zhurnal Strukturnoi Khimii **8**, 548-549 (1967).

4 B. A. Maksimov, V. V. Ilyukhin, Y. A. Kharitonov, and N. V. Belov, Kristallografiya **15**, 926-933 (1970).

5 J. D. Webster, M. E. Westwood, F. H. Hayes, R. J. Day, R. Taylor, A. Duran, M. Aparicio, K. Rebstock, and W. D. Vogel, J. European Ceram. Soc. **18**, 2345-2350 (1998).

6 S. P. Murarka and C. C. Chang, App. Phys. Lett. **37**, 639-641 (1980).

7 F. M. Dheurle, Thin Sol. Films **105**, 285-292 (1983).

8 S. P. Murarka, J. Vac. Sci. Technol. **17**, 775 (1980).

9 O. G. Alekseev, N. V. Krivosheev, M. Y. Khodos, V. R. Galakhov, V. V. Shumilov, V. M. Cherkashenko, E. Z. Kurmaev, and V. A. Gubanov, Inorg. Mater. **22**, 1748-1751 (1986).

10 J. F. Moulder, W. F. Stickle, P. E. Sobol, and K. D. Bomben, *Handbook of X-ray Photoelectron Spectroscopy* (Perkin-Elmer Corporation, Eden Prairie, MN, 1992).

11 Y. Nigara, M. Ishigame, and T. Sakurai, J. Phys. Soc. Japan **30**, 453 (1971).

12 Y. K. Lee, N. Fujimura, and T. Ito, J. Alloys Compd. **193**, 289-291 (1993).

Mat. Res. Soc. Symp. Vol. 611 © 2000 Materials Research Society

Electrical Characteristics of TaOₓNᵧ for High-k MOS Gate Dielectric Applications

Kiju Im, Hyungsuk Jung, Sanghun Jeon, Dooyoung Yang* and Hyunsang Hwang
Department of Materials Science and Engineering, Kwangju Institute of Science and Technology,
#1, Oryong-dong, Puk-gu, Kwangju, 500-712, KOREA
*Jusung Engineering, #49, Neungpyeong, Opo, Kwangju-gun, Kyunggi, 464-890, KOREA

ABSTRACT

In this paper, we report a process for the preparation of high quality amorphous tantalum oxynitride (TaO_xN_y) via ammonia annealing of Ta_2O_5 followed by wet reoxidation for use in gate dielectric applications.   Compared with tantalum oxide(Ta_2O_5), a significant improvement in the dielectric constant was obtained by the ammonia treatment followed by light reoxidation in a wet ambient.   We confirmed nitrogen incorporation in the tantalum oxynitride (TaO_xN_y) by Auger Electron Spectroscopy. By optimizing the nitridation and reoxidation process, we obtained an equivalent oxide thickness of less than 1.6nm and a leakage current of less than $10mA/cm^2$ at -1.5V.   Compared with NH_3 nitridation, nitridation of Ta_2O_5 in ND_3 improve charge trapping and charge-to-breakdown characteristics of tantalum oxynitride.

INTRODUCTION

The scaling of gate dielectric thickness represents the most important issue in the development of the next generation of metal-oxide-semiconductor field effect transistor (MOSFET) devices.   Considering the technology roadmap, an equivalent oxide thickness of less than 1.5nm will be necessary to meet the requirements for sub-100nm MOSFET devices [1]. Due to the low dielectric constant and high tunneling leakage current, the scaling of SiO_2 to the thickness below 2.5nm is impossible.   To satisfy the requirements for sub-100nm MOSFET devices, it will be necessary to develop materials with excellent electrical characteristics, such as a dielectric constant higher than 30, an interface state density of less than $1x10^{11}/cm^2$-eV, a tunneling current of less than $10mA/cm^2$ at the operating bias condition and negligible hysterisis. No alternative high dielectric constant materials, which are capable of meeting the above requirements for sub-100nm MOSFET devices have been reported to date.   Although Ta_2O_5 has been investigated in terms of MOS gate dielectric applications, it is difficult to obtain an equivalent oxide thickness of less than 2nm with acceptable leakage current [2,3].   Since an approximately 1nm-thick interfacial SiO_2 layer is necessary to minimize interface state density and the intermixing of silicon and Ta_2O_5, the dielectric constant of Ta_2O_5 is not sufficient to obtain an equivalent dielectric thickness of less than 2nm. It is known that nitrogen plasma annealing can significantly reduce leakage current and trap density [4]. In addition, it has been reported that the dielectric constant of Ta_2O_5, when deposited on a Ru layer was significantly improved by rapid thermal nitridation [5]. In this paper, we wish to report the preparation of a TaO_xN_y thin film, formed by nitridation followed by wet reoxidation of Ta_2O_5 for gate dielectric applications in sub-100nm MOSFET devices.

EXPERIMENTS

After cleaning a p-type silicon wafer using standard procedure, a 1nm-thick SiO_2 layer was grown by plasma oxidation in order to reduce the interface state density between Si and Ta_2O_5. 8-nm thick Ta_2O_5 films was deposited at 400°C using O_2 and $Ta(OC_2H_5)_5$ source. Rapid thermal nitridation (RTN) was performed in an atmosphere of NH_3 for 30 sec at various temperatures. For comparison, rapid thermal nitridation in ND_3 ambient was also performed. For some samples, an additional wet reoxidation was performed at various temperatures for a 10min. After a 200nm-thick aluminum deposition, MOS devices with a gate area of 9×10^{-6} cm^2 were defined by photolithography and etching.

RESULTS AND DISCUSSION

Figure 1 shows an XRD spectra of as-deposited Ta_2O_5 (8nm) films and processed via rapid thermal nitridation in NH_3 ambient at 700°C and 800°C for 30 seconds. As-deposited and 700°C nitrided Ta_2O_5 were amorphous, while crystalline peaks were observed at 800°C nitrided-Ta_2O_5 film [6].

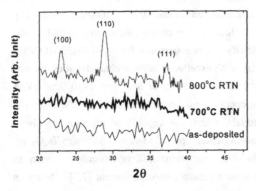

Figure 1 XRD spectra of as-deposited Ta_2O_5 (8nm) films and rapid thermal nitrided in NH_3 ambient at 700°C and 800°C for 30 seconds in NH_3 ambient.

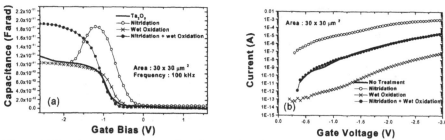

Figure 2 Electrical characteristics of as-deposited Ta_2O_5(8nm), nitrided Ta_2O_5, oxidized Ta_2O_5, and reoxidized-nitrided Ta_2O_5. (a) capacitance versus gate bias measured (b) leakage current versus gate bias

Figure 2 (a) shows the C-V characteristics for as deposited Ta_2O_5, nitrided Ta_2O_5, oxidized Ta_2O_5, and reoxidized-nitrided Ta_2O_5. The accumulation capacitance of as deposited Ta_2O_5 is approximately 11pF, indicating an equivalent oxide thickness of 2.8nm. After nitridation of Ta_2O_5 in NH_3 at 700°C, the accumulation capacitance is approximately 19pF, which indicates an equivalent oxide thickness of 1.6nm. However the capacitance value was degraded when the gate bias was decreased to values below -1.5V, which can be explained by the high leakage current of nitrided Ta_2O_5 as shown Figure 2(b) [7]. To minimize the leakage current of the as-deposited Ta_2O_5, we performed a wet reoxidation of the Ta_2O_5 at 450°C for 10 min. It is reported that high temperature annealing can cause intermixing of Si and Ta_2O_5. We choose a wet reoxidation process instead of a high temperature dry oxidation [8]. After the wet reoxidation of the as-deposited Ta_2O_5 at 450°C for 10 min, the degradation of capacitance is negligible. However, we found a reduction of leakage current of 3~4 orders of magnitude, after wet reoxidation of the as deposited Ta_2O_5 as shown Figure 2(b). To obtain both a high accumulation capacitance and low leakage current, we performed nitridation, followed by a wet reoxidation. The findings show that the accumulation capacitance of reoxidized-nitrided Ta_2O_5 is as high as that of nitrided Ta_2O_5. In addition, the leakage current of the reoxidized-nitrided Ta_2O_5 is comparable to that of the as deposited Ta_2O_5 as shown in Figure 2(b). In other words, a significant improvement in the accumulation capacitance can be obtained without degradation of leakage current.

To evaluate the effect of temperature on device characteristics, the nitridation and reoxidation at various temperatures were performed. The result is shown in Figure 3(a). As respected, the capacitance and leakage current increase with increasing nitridation temperature. In contrast, the capacitance and leakage current decreases with increasing reoxidation

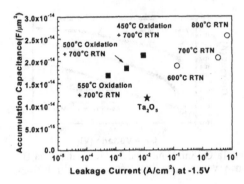

Figure 3 Capacitance versus leakage current for various nitridation and reoxidation conditions. Capacitance and leakage current increases with increasing nitridation temperature. Capacitance and leakage current decreases with increasing reoxidation temperature.

temperature. It was found that the wet reoxidation of nitrided Ta_2O_5 at 450°C for 10min can reduce the leakage current over two order of magnitude with no detectable degradation in accumulation capacitance. Figure 4 shows the nitrogen depth profile of the reoxidized-nitrided Ta_2O_5 which was confirmed by Auger Electron Spectroscopy (AES). As expected, nitrogen incorporation was detected in the case of the nitrided Ta_2O_5. We found that the nitrogen concentration at the interface is higher than that at the bulk.

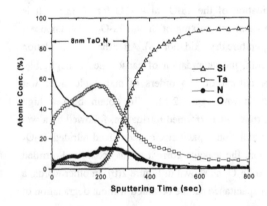

Figure 4 Auger Electron Spectroscopy of 8nm-thick reoxidized-nitrided Ta_2O_5. Nitridation was performed at 700°C for 30 sec and wet reoxidation was performed at 450°C for 10 min.

Based on this, we conclude that the dielectric constant of amorphous TaO_xN_y is significantly higher than that of amorphous Ta_2O_5. It is known that NH_3 annealing of SiO_2 causes a significant charge trapping due to the hydrogen incorporation. We have reported that the charge-trapping of ND_3 nitrided-SiO_2 was significantly less than that of NH_3 annealed samples [9].

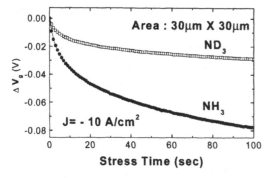

Figure 5 Charge trapping characteristics of TaO_xN_y prepared by NH_3 and ND_3 nitridation under constant current density of $J_g= -10A/cm^2$.

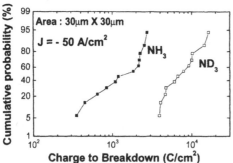

Figure 6 Charge-to-breakdown characteristics under the constant current density of $J_g=-50A/cm^2$. Compared with NH_3 nitridation, charge-to-breakdown characteristics of ND_3 nitrided-TaO_xN_y was significantly improved.

Figure 5 shows that charge trapping characteristics of TaO_xN_y prepared by NH_3 and ND_3 nitridation under the constant current density of $J_g = -10A/cm^2$. Compared with NH_3 nitridation, a significant reduction of charge trapping was observed for ND_3 nitrided-TaO_xN_y. Figure 6 shows that charge-to-breakdown characteristics under the constant current density of $J_g = -50A/cm^2$. Compared with NH_3 nitridation, charge-to-breakdown characteristics of ND_3 nitrided-TaO_xN_y was significantly improved. The larger charge-to-breakdown of ND_3 nitrided-TaO_xN_y is due to reduced charge trapping.

CONCLUSIONS

We have investigated high quality tantalum oxynitride (TaO_xN_y) for use in gate dielectric applications of MOS devices. Nitridation in ammonia ambient increases the dielectric constant of Ta_2O_5 and light reoxidation in a wet ambient reduces the leakage current of the Ta_2O_5. By optimizing the nitridation and reoxidation process, we obtained an equivalent oxide thickness as thin as 1.6nm and a leakage current of less than $10mA/cm^2$ at -1.5V. We also confirmed nitrogen incorporation in the amorphous tantalum oxynitride (TaO_xN_y) by AES. Compared with NH_3 nitridation, nitridation of Ta_2O_5 in ND_3 improves charge trapping and charge-to-breakdown characteristics of tantalum oxynitride. We conclude that TaO_xN_y thin film, formed by nitridation and wet reoxidation of Ta_2O_5 is a promising alternative for future MOS gate dielectric applications.

ACKNOWLEDGMENT

This research was supported by System IC 2010 project, Jusung Engineering Co., Brain Korea 21 Project and Korea Research Foundation.

REFERENCES

1. Semiconductor Industry Association, The International Technology Roadmap for Semiconductor, 1999.
2. Q. Lu, D. Park, A. Kalnitsky, C. Chang, C. C. Cheng, S. P. Tay T. J. King and C. Hu, IEEE Electron Device Lett., 19, 341 (1998).
3. I. C. kizilyalli, R. Y. S. Huang and P. K. Roy, IEEE Electron Device Lett., 19, 423 (1998).
4. G. B. Alers, R. M. Fleming, Y. H. Wong, B. Dennis and A. Pinczuk, G Redinbo, R. Urdahl, E. ong, and Z. Hasan, Appl. Phys. Lett., 72, 1308 (1998).
5. J. Lin, N. Massaki, A. Tsukune and M. Yamada, Appl. Phys. Lett., 74, 2370 (1999).
6. S. W. Park, Y. K. Baek, J. Y. Lee, C.O. Park, and H. B. Im, J.Electron Mat. Vol.21 No.6, 1992
7. W. K. Henson, K. Z. Ahmed, E. M. Vogel, J. R. Hauser, J. J. Wortman, R. D. Venables, M. Xu, and D. Venable, IEEE Electron Device Lett., 20, 179 (1999).
8. G. B. Alers, D. J. Werder, Y. Chabal, H. C. Lu, E. P. Gusev, E. Garfunkel, T. Gustafsson, and R. S. Urdahl, Appl. Phys. Lett., 73, 1517 (1998).
9. H. Kwon and H. Hwang, Appl. Phys. Lett., 76, 772 (2000).

Novel Gate Insulators

Mat. Res. Soc. Symp. Vol. 611 © 2000 Materials Research Society

Ultra-thin Gate Oxide Prepared by Nitridation in ND$_3$ for MOS Device Applications

Hyungshin Kwon and Hyunsang Hwang
Department of Materials Science and Engineering
Kwangju Institute of Science and Technology
1, Oryong-dong, Puk-gu, Kwangju, 500-712, KOREA, email: hwanghs@kjist.ac.kr

ABSTRACT

The electrical and reliability characteristics of ultra-thin gate oxide, annealed in ND$_3$ gas, have been investigated. Compared with a control oxide, which had been annealed in NH$_3$, the ND$_3$-nitrided oxide exhibits a significant reduction in charge trapping and interface state generation. The improvement of electrical and reliability characteristics can be explained by the strong Si-D bond at the Si/SiO$_2$ interface. This nitridation process of gate dielectric using ND$_3$ has considerable potential for future ultra large scaled integration (ULSI) device applications.

INTRODUCTION

As device sizes are scaled down in order to achieve ultra large scaled integration (ULSI), the reliability of ultra-thin gate oxide becomes one of the most critical factors in metal-oxide-semiconductor field effect transistor (MOSFET) technology. The proper scaling of gate oxide thickness can improve current driving capability and reduce short channel effects. However, ultra-thin oxide films have numerous reliability problems, including time-dependent dielectric breakdown, interface state generation, charge trapping, and threshold voltage shifts as the results of penetration of dopant.

Nitridation of the gate oxide in NH$_3$ has been investigated in order to improve gate dielectrics integrity [1]. The incorporation of nitrogen in gate dielectrics leads to reduced boron penetration which is a contributor to threshold voltage instability, especially for ultra-thin gate oxide with a thickness of less than 5nm [2]. However, post-oxidation annealing in NH$_3$ requires an additional reoxidation to reduce the degradation of dielectric reliability [1]. Without this reoxidation, the electron trapping characteristics of NH$_3$ nitrided gate oxide is significantly high, because of the high concentration of hydrogen. It is known that the electrical and reliability characteristics of the dielectric are dependent on the concentration of hydrogen-related species in the oxide bulk and the interface [3, 4]. According to Yoshii et al., forming gas annealing at high temperatures significantly enhances both interface and oxide trap generation under electrical stress [5].

According to Lyding et al., a significant improvement in the reliability characteristics under hot carrier stress was observed, after the deuterium annealing of a MOSFET device

[6-8]. Because of the heavy mass of deuterium, Si-D bonds are more difficult to break than Si-H bonds under hot carrier stress [7]. However, no improvement in MOSFET hot carrier reliability was observed when the annealing was performed after the deposition of a silicon nitride layer [7, 8]. Since the silicon nitride is a strong barrier to the diffusion of deuterium and hydrogen, it is difficult to incorporate a sufficient amount of deuterium at the Si/SiO$_2$ interface by means of deuterium annealing temperatures of about 450 °C after metallization.

We recently reported a new gate oxidation process using D$_2$O (deuterium oxide) as the oxidizing gas [9]. Compared with conventional wet oxidation using H$_2$O, the gate oxide which was grown in D$_2$O ambient exhibits a significant improvement of device reliability under electrical stress. In this paper, we present a new method for incorporating deuterium and nitrogen at the Si/SiO$_2$ interface using post-oxidation annealing in ND$_3$.

EXPERIMENTAL DETAILS

Conventional polysilicon gate MOS capacitors, with gate oxides which are 6.0 nm in thickness, were fabricated on n-type silicon wafers. The gate oxide was grown in an oxygen ambient, followed by nitrogen annealing at 980 °C for 30min. Nitridation in an ND$_3$ ambient was performed at 800 °C for 30, 45, and 60 min. The process pressure was 0.1atm. For comparison, nitridation in NH$_3$ was also carried out at 770, 800, and 830 °C. After nitridation of the gate oxide, a 200nm-thick n$^+$ in-situ doped polycrystalline silicon layer was deposited. Various devices with different gate areas were defined by photolithography and etching.

RESULTS AND DISCUSSIONS

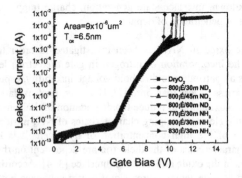

Fig. 1 Current versus voltage characteristics of MOS capacitors with 6.5nm-thick dielectrics grown in various process conditions.

Figure 1 shows current versus voltage characteristics of MOS capacitors. Considering the same tunneling current of various devices, we were able to confirm that the thickness of the gate oxide was nearly the same for various nitridation conditions.

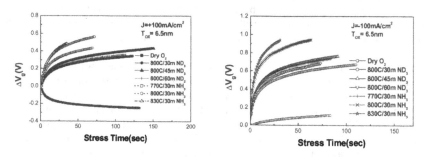

Fig. 2 Gate voltage shift under constant current stress for both samples. The gate voltage shift under constant current stress is due to the charge trapping in the oxide. Compared with NH$_3$ annealing, the MOS capacitors with an oxide annealed in an ND$_3$ ambient exhibit less electron trapping.

Fig. 2 shows the gate voltage shift under constant current stress for both samples. This voltage shift is due to the trapping of electrons in the oxide. With increasing nitridation temperature in the NH$_3$ ambient, a significant increase in electron trapping was observed. Compared with NH$_3$ annealing, the MOS capacitors with oxide annealed in an ND$_3$ ambient exhibited a lower level of electron trapping. It is known that hydrogen species cause electron trapping under constant current stress. The reduced charge trapping under constant current stress can be explained by the large deuterium kinetic isotope effect [7].

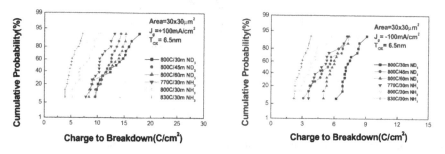

Fig. 3 Charge-to-breakdown characteristics under constant current stress. Compared with oxide annealed in NH$_3$, the charge-to-breakdown of gate oxide annealed in ND$_3$ ambient is significantly improved.

Fig. 3 shows charge-to-breakdown characteristics under constant current stress. In order to investigate the intrinsic characteristics of the gate oxide, small area (A = $9 \times 10^{-6} cm^2$) capacitors were measured. Compared with oxide which had been annealed in NH_3, the charge-to-breakdown of the gate oxide annealed in the ND_3 ambient was significantly improved. Since it is generally accepted that the breakdown mechanism of the gate dielectric is related to charge trapping within the dielectric, the larger charge-to-breakdown of the oxide annealed in ND_3 must be due to reduced charge trapping.

To investigate the endurance characteristics of the interface state, we measured low-frequency capacitance versus gate voltage under electrical stress, as shown in Fig. 4. The initial C-V curves were nearly the same for various nitridation conditions which indicate a similar initial interface state density. Compared with the NH_3 annealed samples, the distortion of the C-V curves under electrical stress were significantly reduced for the case of the gate oxide which had been annealed in the ND_3 ambient. In other words, the oxide which had been annealed in ND_3 generated a smaller amount of interface state density under the same amount of electrical stress. In the case of the oxide which had been annealed in ND_3, we believe that the dangling bonds at the Si/SiO_2 interface were rendered passive by deuterium. Considering the strong bond force for Si-D, we can explain a lesser generation of interface state under the same amount of electrical stress.

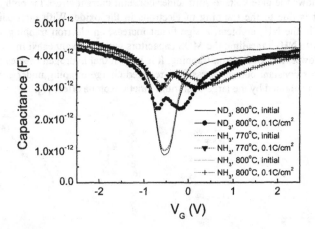

Fig. 4 Low-frequency capacitance versus gate voltage under electrical stress. Compared with NH_3 annealed samples, the distortion of C-V curves under the electrical stress were significantly reduced for the gate oxide annealed in ND_3 ambient.

CONCLUSIONS

We investigated a new nitridation process using ND_3. Compared with post-oxidation annealing in NH_3, annealing in an ND_3 ambient improves the reliability characteristics of the gate oxide under electrical stress. Gate oxides which had been annealed in an ND_3 ambient exhibited less charge trapping, less generation of interfaces state, and a larger charge-to-breakdown under electrical stress. The fact that Si-D bonds are more difficult to break than Si-H bonds due to the heavy mass of deuterium, explains the improvement in the gate oxide which had been annealed in ND_3. The use of ND_3 as an annealing medium for gate dielectric appears to have considerable potential.

ACKNOWLEDGMENT

This research was supported by Hyundai Electronics Industries Company, Brain Korea 21 Project and Korea Research Foundation.

REFERENCES

1. T. Hori, H. Iwasaki, and K. Tsuji, IEEE Trans. on Electron Device, **36**, 340 (1989).
2. H. Hwang, W. Ting, D. L. Kwong, and J. Lee, Proc. IEEE International Electron Device Meeting **1990**, 421 (1990)
3. J. Yugami, T. Itoga, and M. Ohkura, Proc. IEEE International Electron Device Meeting **1995**, 855 (1995).
4. Y. N. Cohen and T. Gorczyca, IEEE Electron Device Lett. **9**, 287 (1988).
5. I. Yoshii, K. Hama and K. Hashimoto, IEEE Proc. International Reliability Physics Symposim **1992**, 136 (1992).
6. J. W. Lyding, K. Hess and I. C. Kizilyalli, Appl. Phys. Lett., **68**, 2526 (1996).
7. I. C. Kizilyalli, J. W. Lyding and K. Hess, IEEE Electron Device Lett. **18**, 81 (1997).
8. J. Lee, S. Aur, R. Eklund, K. Hess and J. W. Lyding, Journal of Vac. Sci. Technol. A **16(3)**, 1762 (1998).
9. H. Kim and H. Hwang, Appl. Phys. Lett. **74**, p. 709 (1999).

Mat. Res. Soc. Symp. Vol. 611 © 2000 Materials Research Society

Amorphous Mixed TiO$_2$ and SiO$_2$ Films on Si(100) by Chemical Vapor Deposition

Ryan C. Smith [a], Charles J. Taylor [a], Jeffrey Roberts [a], Noel Hoilien[b], Stephen A. Campbell [b], and Wayne L. Gladfelter [a,*]

[a] Departments of Chemistry and [b] Computer and Electrical Engineering, University of Minnesota, Minneapolis, MN 55455.

ABSTRACT

Amorphous thin films of composition Ti$_x$Si$_{1-x}$O$_2$ have been grown by low pressure chemical vapor deposition on silicon (100) substrates using Si(O-Et)$_4$ and either Ti(O-^{i}Pr)$_4$ or anhydrous Ti(NO$_3$)$_4$ as the sources of SiO$_2$ and TiO$_2$, respectively. The substrate temperature was varied between 300 and 535°C, and the precursor flow rates ranged from 5 to 100 sccm. Under these conditions growth rates ranging from 0.6 to 90.0 nm/min were observed. As-deposited films were amorphous to X-rays and SEM micrographs showed smooth, featureless film surfaces. Cross-sectional TEM showed no compositional inhomogeneity. RBS revealed that x (from the formula Ti$_x$Si$_{1-x}$O$_2$) was dependent upon the choice of TiO$_2$ precursor. For films grown using TTIP-TEOS x could be varied by systematic variation of the deposition conditions. For the case of TN-TEOS x remained close to 0.5 under all conditions studied. One explanation is the existence of a specific chemical reaction between TN and TEOS prior to film deposition. TEOS was mixed with a CCl$_4$ solution of TN at room temperature to produce an amorphous white powder (Ti/Si = 1.09) and ^{1}HNMR of the CCl$_4$ solution indicated resonances attributable to ethyl nitrate.

INTRODUCTION

As the thickness of the gate oxide in field effect transistors approaches 2 nm, direct tunneling will lead to unacceptable leakage currents making it necessary to replace the silicon dioxide layer with a material possessing a higher dielectric constant. In our investigations into the chemical vapor deposition of TiO$_2$ (κ = 19-30) polycrystalline films were produced.[1-3] For the thin films necessary in microelectronics, surface roughness and grain boundaries resulting from polycrystalline films can have an undesirable effect on properties such as the dielectric breakdown and leakage current. The high κ material can be made amorphous, thus eliminating grain boundaries, by combining it with a low κ, glass forming oxide such as SiO$_2$. Titanium silicon oxide films have been produced by sol-gel,[4, 5] flame hydrolysis,[6, 7] and plasma enhanced CVD[8, 9] for optical applications and some reports of their electrical properties have also appeared.[8, 10] In this paper we describe the deposition of titanium silicon oxide films onto Si(100) by the simultaneous chemical vapor deposition of TiO$_2$ and SiO$_2$ and note the dramatic effect that the precursor structure plays in determining the film composition.

EXPERIMENTAL DETAILS

Tetraethylorthosilicate (TEOS) was used as the source of silicon dioxide while titanium tetraisopropoxide (TTIP) or anhydrous titanium tetranitrate (TN) was used to deposit the TiO$_2$ component. Precursors were purchased from Aldrich, and TEOS and TTIP were distilled under vacuum prior to use. All precursors were stored and the vessels were loaded under an inert N$_2$

atmosphere. Substrates were cleaned using methylene chloride followed by a 7:3 solution of H_2SO_4 and 30% H_2O_2. Prior to deposition a wafer was removed from this solution, rinsed thoroughly with deionized water, dipped into a 10% HF solution for 15 seconds, and blown dry with compressed air. The wafer was then immediately placed into a cold wall LPCVD reactor, which was evacuated using a mechanical pump. The TEOS precursor vessel was cooled to 0°C in an ice bath while the TN and TTIP vessels were warmed to 40 and 37°C, respectively. The high purity N_2 flow rate was varied from 5 to 50 sccm through the TEOS vessel and from 5 to 100 sccm through the TN and TTIP vessels. Overall reactor pressures of 0.5 to 1.5 Torr were maintained during the process. The molybdenum susceptor was heated to the desired reaction temperature (300-535°C) using a Variac-controlled 1000 W halogen lamp set within a parabolic, polished aluminum reflector. The TEOS vessel was opened first and allowed to equilibrate, then the TN or TTIP vessel was opened for the duration of the deposition.

DISCUSSION

Under all conditions studied no film growth was observed using TEOS alone. This was consistent with minimum temperatures of 650°C reported for the thermal CVD of SiO_2[11, 12] and with TPD studies showing no evidence of TEOS thermal decomposition on SiO_2 powders.[13] As TN or TTIP was introduced along with the TEOS, film growth of the mixed oxide was observed at rates ranging from 0.6 to 90 nm/min. Interruption of the TN or TTIP delivery (while maintaining the flow of TEOS) halted all film growth. After delivery of TEOS was stopped pure TiO_2 deposition continued, as long as TN or TTIP flow was continued.

The as-deposited films were amorphous to X-rays and plan-view SEM images showed flat, featureless surfaces. However, after annealing at 600°C under N_2 for 2.5 hours reflections corresponding to anatase TiO_2 were observed in the XRD spectrum of films grown from the TTIP-TEOS precursor pair. Cross-sectional TEM of as-deposited films showed no crystalline structure or evidence of compositional inhomogeneity within the mixed oxide films. A 1.4-1.8 nm interfacial layer was observed between silicon and the $Ti_xSi_{1-x}O_2$ film. This layer has been observed in other studies in which TiO_2 has been deposited on silicon.[2, 3, 14] The formation of the layer could be due to exposure of the substrate to impurities in the gas at the beginning of the deposition or to an expected reaction between TiO_2 and Si.[15] Rutherford backscattering spectrometry confirmed the presence of both Ti and Si throughout the films. The spectra were modeled by compositions that corresponded to a mixture of titanium and silicon dioxide, $Ti_xSi_{1-x}O_2$. Figure 1 shows the RBS spectrum for a film grown using TEOS and TTIP in which the TEOS vessel was closed after approximately 300 nm of mixed oxide was deposited. The top layer corresponded to pure TiO_2 and the transition between the pure and mixed phases was sharp relative to the depth resolution of the RBS experiment.

Using the TTIP-TEOS precursor pair the ratio of TiO_2 to SiO_2 was controlled by systematic variation of the carrier gas flow rate through the precursor vessels. For a growth temperature of 500°C and a constant TEOS flow rate (10 sccm), figure 2 shows that x ($Ti_xSi_{1-x}O_2$) increased as the flow rate through the TTIP vessel increased from 5–100 sccm. Conversely, increasing the flow rate through the TEOS while maintaining a constant rate through the TTIP (20 sccm) caused x to decrease. The effect seemed to saturate at higher flow rates and more research is necessary to establish whether the shape of these curves was related to changes in transport phenomena or to deposition chemistry. While an independent measurement of the

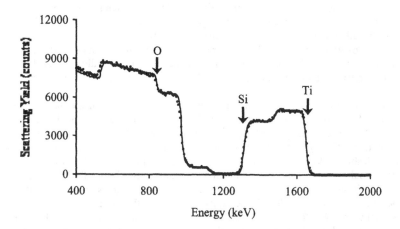

Figure 1. *RBS of film grown at 500°C using TEOS (N_2 flow = 25 sccm) and TTIP (N_2 flow = 20 sccm). TEOS vessel was closed after approximately 300 nm of mixed oxide had grown. The model corresponds to 310 nm of pure TiO_2 on top of 290 nm of $Ti_{0.76}Si_{0.24}O_2$.*

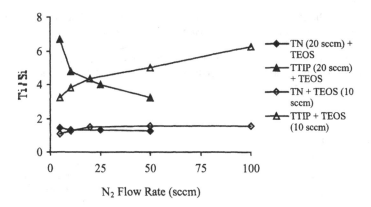

Figure 2. *Variation of film composition (ratio of Ti to Si) as a function of precursor delivery rate.*

precursor partial pressures was not obtained, upper estimates based on the equilibrium vapor pressures (0.018 Torr for TTIP at 37°C and 0.28 Torr for TEOS at 0°C)[16] suggested that TEOS was always present in excess.

In striking contrast to the behavior of the TTIP-TEOS reactants, similar variation in carrier gas flow rate produced virtually no change in film stoichiometry (x) in the TN-TEOS system. Under all conditions studied, x remained close to 0.5.

The deposition rate increased with increasing temperature for both sets of precursors. In addition the composition of films grown using TTIP-TEOS monotonically decreased as temperature increased (figure 3). For the TN-TEOS system, however, x remained constant at all temperatures.

The binary phase diagram for the TiO_2-SiO_2 system does not indicate the presence of any unique phases.[17, 18] At 500°C limited solubility of one component in the other is observed, therefore the films produced in this study do not represent equilibrium compositions. In the TTIP-TEOS deposition system, the compositional variation reflects the deposition conditions in a systematic fashion. When TN is used as the TiO_2 precursor this is not the case. A plausible explanation would involve a specific chemical reaction between TN and TEOS during the early stages of deposition (either in the gas phase or on the surface). Studies of the reactivity of anhydrous metal nitrates, including TN, established them as effective nitrating reagents, e. g. alcohols react with TN to form organonitrates.[19, 20] The reaction of TEOS with TN proposed in eq. 1, is very similar to the reaction of ROH with TN, and could explain the one-to-one composition of films grown with this combination of precursors.

$$Si(OEt)_4 \ + \ Ti(NO_3)_4 \longrightarrow \ TiO_2/SiO_2 \ + \ 4 \ EtONO_2 \qquad (1)$$

The thermal instability of ethyl nitrate may preclude its observation during deposition at the temperatures used in this study.

To explore the efficacy of this reaction $Si(OEt)_4$ was mixed with a CCl_4 solution of $Ti(NO_3)_4$. After stirring for 30 min at room temperature an amorphous white powder formed. ICP-MS analysis of the isolated powder gave a ratio of titanium to silicon of 1.09. Also, ^{1}HNMR spectroscopy of the CCl_4 solution indicated the formation of resonances at 4.52 and 1.40 ppm attributable to ethyl nitrate. In addition to its relevance to the CVD process, eq. 1 may have value as an alternative, low temperature synthesis of $Ti_xSi_{1-x}O_2$ powders.

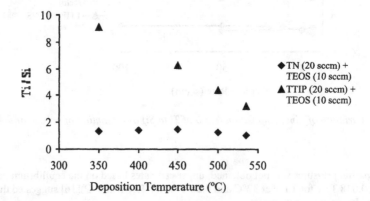

Figure 3. *Variation of film composition (ratio of Ti to Si) with deposition temperature.*

ACKNOWLEDGEMENTS

The authors would like to thank Drs. Christopher Hobbs, Mike Tiner and Rama Hegde for discussions and TEM work. This research was supported by grants from the Semiconductor Research Corporation and the National Science Foundation (CHE - 9616501).

REFERENCES

[1]   S. A. Campbell, D. C. Gilmer, X.-C. Wang, M.-T. Hsieh, H.-S. Kim, W. L. Gladfelter, and J. Yan, *IEEE Trans. Electron Devices*, vol. 44, pp. 104 - 109, 1997.

[2]   D. C. Gilmer, D. G. Colombo, C. J. Taylor, J. Roberts, G. Haugstad, S. A. Campbell, H.-S. Kim, G. D. Wilk, M. A. Gribelyuk, and W. L. Gladfelter, *Chem. Vap. Deposition*, vol. 4, pp. 9 - 11, 1998.

[3]   C. J. Taylor, D. C. Gilmer, D. G. Colombo, G. D. Wilk, S. A. Campbell, J. Roberts, and W. L. Gladfelter, *J. Am. Chem. Soc.*, vol. 121, pp. 5220 - 5229, 1999.

[4]   S. M. Melponder, A. W. West, C. L. Barnes, and T. N. Blanton, *J. Mater. Sci.*, vol. 26, pp. 3585-3592, 1991.

[5]   R. R. A. Syms and A. S. Holmes, *J. Non-Cryst. Sol.*, vol. 170, pp. 223-233, 1994.

[6]   P. C. Schultz, *J. Am. Ceram. Soc.*, vol. 59, pp. 214-219, 1976.

[7]   S. M. Mukhopadhyay and S. H. Garofalini, *J. Non-Cryst. Sol.*, vol. 126, pp. 202-208, 1990.

[8]   T. Kamada, M. Kitagawa, M. Shibuya, and T. Hirao, *Jpn. J. Appl. Phys.*, vol. 30, pp. 3594-3596, 1991.

[9]   C. Martinet, V. Paillard, A. Gagnaire, and J. Joseph, *J. Non-Crystalline Solids*, vol. 216, pp. 77-82, 1997.

[10]   M. Inoue, U. S. Patent 3 614 548, 1971.

[11]   A. C. Adams and C. D. Capio, *J. Electrochem. Soc.*, vol. 126, pp. 1042-1046, 1979.

[12]   S. Rojas, A. Modelli, W. S. Wu, A. Borghesi, and B. Pivac, *J. Vac. Sci. Technol. B*, vol. 8, pp. 1177-1184, 1990.

[13]   T. Okuhara and J. M. White, *Appl. Surf. Sci.*, vol. 29, pp. 223-241, 1987.

[14]   N. Rausch and E. P. Burte, *J. Electrochem. Soc.*, vol. 140, pp. 145 - 149, 1993.

[15]   K. J. Hubbard and D. G. Schlom, *J. Mater. Res.*, vol. 11, pp. 2757 - 2776, 1996.

[16]   W. G. Mallard and P. J. Linstrom, "NIST Chemistry Webbook, NIST Standard Reference Database Number 69," . Gaithersburg, MD 20899: National Institute of Standards and Technology, February 2000, (http://webbook.nist.gov).

[17]   R. C. DeVries, R. Roy, and E. F. Osborn, *Transactions of the British Ceramic Society*, pp. 525-540, 1954.

[18]   D. L. Evans, *J. Am. Ceram. Soc.*, vol. 53, pp. 418-419, 1970.

[19]   C. C. Addison and N. Logan, *Adv. Inorg. Chem. and Radiochem.*, vol. 6, pp. 71 - 142, 1964.

[20]   C. C. Addison and W. B. Simpson, *J. Chem. Soc.*, pp. 598 - 602, 1965.

Mat. Res. Soc. Symp. Vol. 611 © 2000 Materials Research Society

Zr OXIDE BASED GATE DIELECTRICS WITH EQUIVALENT SiO$_2$ THICKNESS OF LESS THAN 1.0 nm AND DEVICE INTEGRATION WITH Pt GATE ELECTRODE

Yanjun Ma and Yoshi Ono
Sharp Laboratories of America, 5700 NW Pacific Rim Blvd, Camas, WA 98607

ABSTRACT

ZrO$_2$ films are investigated as an alternative to SiO$_2$ gate dielectric below 1.5nm. A maximum accumulation capacitance ~35 fF/μm^2 with a leakage current of less than 0.1 A/cm^2 has been achieved for a 3 nm Zr-O film, suggesting that ZrO$_2$ can be scaled to below an equivalent oxide thickness of 0.5 nm. Al and Si doping is also investigated to reduce leakage currents and to increase the crystallization temperature of the film. Submicron MOSFETs with TiN or Pt gate electrodes have been fabricated with these gate dielectrics with excellent characteristics, demonstrating the feasibility of CMOS process integration. In particular, Pt damascene gate PMOS is shown to have the proper threshold voltage for dual metal gate CMOS application.

INTRODUCTION

With shrinking device feature sizes, the semiconductor industry is facing serious challenges on many technology fronts. One of the grand challenges is the replacement of the SiO$_2$/poly-Si gate stack with new gate dielectrics and gate electrodes. There are many material choices in the search for a replacement for SiO$_2$, including TiO$_2$, Al$_2$O$_3$ and Ta$_2$O$_5$, and more recently, Zr and Hf oxides and silicates [1-10]. In this work we report on the investigation of zirconium oxide dielectrics which showed promise for use to equivalent oxide thickness (EOT) < 1.0 nm. In particular, we concentrate on the scaling limit of ZrO$_2$. This information will greatly aid in the selection of a SiO$_2$ replacement. A maximum accumulation capacitance about 35 fF/μm^2 with a leakage current less than 0.1 A/cm^2 has been achieved for a 3 nm Zr-O film. This puts the scaling limit of ZrO$_2$ to an EOT of below 0.5 nm. Al doping (~25% by XPS) is used to increase the crystallization temperature of the gate dielectrics. Submicron MOSFETs with these gate dielectrics have been fabricated with a nitride replacement process. Excellent characteristics have been achieved, demonstrating the feasibility of integrating this gate dielectric in a CMOS process flow.

Together with new high k gate dielectrics, dual metal gate schemes may be needed within the next few years to enable the continued scaling of CMOS devices. Low work function metals are needed to replace n+ polysilicon for NMOSFETs and high work function metals are needed to replace p+ polysilicon in PMOSFETs. While there are many materials with the right work function for NMOS, including Al, Ta, TaN, Mo, and Zr, there are few materials that have an appropriate work function for the gate electrode of the PMOSFET. Pt has one of the highest work functions for pure metals and is a strong candidate for the gate electrode of PMOS metal gate device. In this work we also demonstrate the feasibility of fabricating Pt damascene gate and integrating it with a ZrO$_2$

based high κ gate dielectric in PMOS transistors. Our results show that the Pt gate PMOS has the right threshold voltage for future sub-0.1μm CMOS applications.

SCALING LIMIT OF ZrO₂ BASED DIELECTRICS

ZrO_2 films were prepared by sputtering of a Zr target in a mixture of oxygen and Ar at room temperature. Post deposition anneals at 400~500°C are sufficient to achieve reduced leakage current. This is drastically different than TiO_2 where annealing at more than 750°C is needed to reduce the leakage current.[7] Thickness of the films is evaluated by spectroscopic ellipsometry assuming a single dielectric layer. Doping of Al was done with co-sputtering of an Al target. X-ray results indicate that Al doping prevented the crystallization of the Zr-Al-O film after anneals at up to 800°C. An Al/TiN top electrode was then deposited by sputtering and patterned by dry etch to make 100x100 μm² capacitors for electrical testing.

Figure 1 shows the high frequency CV curves of MOSCAPs with Zr-O films of three different thickness. As can be seen a maximum accumulation capacitance of 35 fF/μm² is obtained for a 3.1nm film at gate bias of -1.5V. Beyond the gate bias of −1.5V the gate leakage current causes the CV curve to rise rapidly, preventing an accurate measurement of the capacitance. By extrapolation it can be estimated that C~38 fF/μm² at gate bias of -2 V. This corresponds to a classical dielectric thickness (CDT = ε_{SiO2}/C) of 0.9 nm. Including the quantum mechanical corrections which has been estimated to be typically about 0.3~0.4 nm, and as high as 0.6-0.7 nm[11,12], an EOT less than 0.5~0.6nm is obtained. Using the 0.3nm quantum correction, we obtained the effective dielectric constant of 20 for the thinnest film (3.1nm) and 17 for the thicker film (4.5 nm).

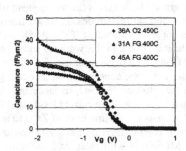

Figure 1. High frequency (100KHz) CV results of ZrO_2 films.

The effective dielectric constant of the film is also dependent on the annealing condition. Using the data in Figure 1, the film which has undergone oxygen annealing at 450°C has an effective dielectric constant of only 15.6. This can be explained by the presence of a thicker interfacial layer at the ZrO_2/Si interface. Thus higher temperature annealing in an oxygen ambient is to be avoided to achieve lower EOTs.

Transmission electron microscope investigation indicates that even with this low temperature anneal there is an interfacial layer present at the ZrO_2/Si interface. From the overall effective dielectric constant, it appears that this interfacial layer is some kind of silicate with dielectric constant higher than that of silicon oxide.

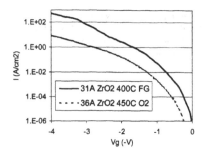

Figure 2. Leakage current of two of ZrO2 films shown in Figure 1.

Figure 2 shows the IV curves for films listed in Figure 1. For the 3.1nm film, the gate leakage current is only ~0.1 A/cm^2 at the likely operating voltage of -1 V. This leakage current is less than the suggested 1~100 A/cm^2 limit for the gate dielectric in CMOS circuits, indicating that ZrO_2 with EOT of less than 0.5nm is achievable.

Our X-ray data shows that ZrO_2 films became crystalline after anneals at temperatures as low as 450°C. By doping with Al, the resulting film remains amorphous after anneals at temperature as high as 800°C. Doping with Al is attractive because Al_2O_3 also has good interface properties with Si.

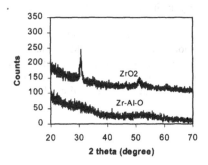

Figure 3. X-ray diffraction results of undoped and Al (~25%) doped ZrO_2 films after 800°C anneal.

One drawback of Al doping is the reduction in the effective dielectric constant. CV results shown in Figure 4 suggest that for films of similar thickness, a reduction of about 25% results from Al doping. The trade-off needs to be considered when choosing between using doped and un-doped ZrO_2 films.

METAL GATE MOSFET FABRICATED WITH ZrO_2 BASED GATE DIELECTRIC

PMOS transistors with ZrO_2 gate dielectrics and metal gates were processed using a nitride gate replacement process [13]. In this process the source/drain are formed with a silicon nitride dummy gate in place of the real gate. After an oxide dielectric fill and chemical mechanical polishing to expose the nitride, the dummy gate is selectively removed to expose the channel region. The gate dielectric is then deposited and is

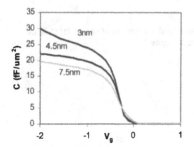

Figure 4. High frequency CV results of 25% Al doped ZrO_2 films. A ~25% reduction in the total capacitance is noted when compared with undoped films shown in Figure 1.

annealed at 400°C for 60s in either O_2 or forming gas. No threshold voltage adjustment implant is performed. From the deposition time, the thickness of the film is estimated to be 6 nm. From capacitance measurement, an accumulation capacitance (at 1.5V) of more than 20 $fF/\mu m^2$ is obtained, yielding a CDT~1.7nm and an EOT~1.4nm. TiN and Pt gate electrodes were sputter-deposited. Devices containing Pt were patterned using a proprietary CMP process. Typical Pt removal rates were 180nm/min at a down force of 4 psi. Polish uniformity was about 5% and oxide erosion is minimal. The final Pt gate electrode thickness is ~200nm. The maximum process temperature after gate stack formation was a 450°C forming gas anneal. This gate replacement process is ideally suited for the fabrication of metal gate devices because of the effect of high temperature treatment on the high k gate dielectrics, particularly the interface with the metal electrode as well as the interface with the Si substrate.

Well-behaved transistor characteristics are obtained on transistors with both Pt and TiN gate and Zr-Al-O gate dielectrics. Sub-threshold behaviors of the PMOS transistors are shown in Figure 5. The threshold voltage is -0.42V for PMOSFETs with Pt/TiN electrode and -0.12V for Pt electrode. The subthreshold swing is ~75 mV/dec. The lower threshold voltage value for the Pt PMOSFET suggests that this is a good candidate for future dual metal gate CMOS.

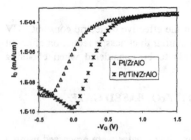

Figure 5, Subthreshold (V_d=-0.1V) of PMOSFET with Pt and TiN gate electrode and Zr-O gate dielectric. Transistor size is 10μmx10μm.

The drain characteristics of Pt gated PMOSFET is shown in Figure 6. Good performance is obtained despite the fact that in our process the high k gate dielectric is sputter-deposited into a trench after the removal of the dummy gate and that there is substantial amount of charge trapping in the dielectric.[10] Clearly chemical vapor deposition of the high k dielectric is preferred since CVD normally has better step coverage than the PVD method used in this work. We are currently investigating such an approach.

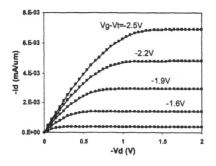

Figure 6. Drain characteristics of PMOSFET with Pt gate electrode and Zr-Al-O gate dielectric.

CONCLUSIONS

In summary, Zr oxide based films are shown to be a promising candidate for replacing SiO₂ as the gate dielectric below 1.5nm. We believe that it can be used for CMOS applications with EOT < 0.5nm. MOSFETs using these high-k gate dielectrics have been fabricated with a gate replacement process and showed satisfactory performance. In particular, we demonstrated the fabrication of PMOSFET with Pt damascene gate and Zr oxide. This type of device has the potential to be used in future dual metal gate CMOS devices.

ACKNOWLEDGEMENTS

We would like to thank David Evans and Lisa Stecker for help with wafer processing.

REFERENCES

[1] H.S. Kim, D.C. Gilmer, S.A. Campbell, and D.L. Polla, Appl. Phys. Lett. 69, 3860 (1996); S.A. Campbell, D.C. Gilmer, X. Wang, M. Hsieh, H. Kim, W. Gladfelter, and J. Yan, IEEE Trans. Electron Dev. 44, 104 (1997).
[2] S.C. Sun and T.F. Chen, Jpn J. Appl. Phys. 36, 1346 (1997).
[3] Y. Momiyama, et al, Symp. VLSI Technology, p135 (1997).
[4] D. Park, Q. Lu, T. King, C. Hu, A. Kalnitsky, S. Tay, C. Cheng, IEDM Tech. Digest, p381 (1998).
[5] X. Guo, T.P. Ma, T. Tamagawa, B.L. Halpern, IEDM Tech. Digest, p377 (1998).
[6] H. F. Luan, B.Z. Wu, L.G. Kang, B.Y. Kim, R. Vrtis, D. Roberts, D.L. Kwong, IEDM Tech. Digest, p609 (1998).
[7] Y. Ma, Y. Ono, and S.T. Hsu, in MRS Proceedings 567, 31 (1999).
[8] G.D. Wilk and R.M. Wallace, Appl. Phys. Lett. 74, 2854, 1999; ibid 76, 112, (2000).
[9] Y. Ma, Y. Ono, L.S. Stecker, D.R. Evans, and S.T. Hsu, IEDM Tech. Dig. pp149-152 (1999).

[10] B. Lee, L. Kang, R. Nieh, Q. Qi, and J. C. Lee, Appl. Phys. Lett. 76, 1926 (2000).

[11] K.S. Krisch, J.D. Bude, and L. Manchanda, IEEE Electron Dev. Lett. 17, 521 (1996).

[12] S. Takagi, M. Takayanagi, and A. Toriumi, , IEEE Trans. Electron Dev. 46, 1446 (1999).

[13] Y. Ma, D.R. Evans, T. Nguyen, Y. Ono, and S.T. Hsu, IEEE EDL 20, 251 (1999).

Novel Gate Structures

Mat. Res. Soc. Symp. Vol. 611 © 2000 Materials Research Society

MOLYBDENUM AS A GATE ELECTRODE FOR DEEP SUB-MICRON CMOS TECHNOLOGY

Pushkar Ranade[†], Yee-Chia Yeo, Qiang Lu, Hideki Takeuchi, Tsu-Jae King, Chenming Hu
Department of Electrical Engineering and Computer Sciences
[†] Department of Materials Science and Engineering
University of California at Berkeley, Berkeley, CA 94720

ABSTRACT:

Molybdenum has several properties that make it attractive as a CMOS gate electrode material. The high melting point (~2610°C) and low coefficient of thermal expansion (5×10^{-6}/ °C, at 20 °C) are well suited to withstand the thermal processing budgets normally encountered in a CMOS fabrication process. Mo is among the most conductive refractory metals and provides a significant reduction in gate resistance as compared with doped polysilicon. Mo is also stable in contact with SiO_2 at elevated temperatures. In order to minimize short-channel effects in bulk CMOS devices, the gate electrodes must have work functions that correspond to E_c (NMOS) and E_v (PMOS) in Si. This would normally require the use of two metals with work functions differing by about 1V on the same wafer and introduce complexities associated with selective deposition and/or etching. In this paper, the dependence of the work function of Mo on deposition and annealing conditions is investigated. Preliminary results indicate that the work function of Mo can be varied over the range of 4.0-5.0V by a combination of suitable post-deposition implantation and annealing schemes. Mo is thus a promising candidate to replace polysilicon gates in deep sub-micron CMOS technology. Processing sequences which might allow the work function of Mo to be stabilized on either end of the Si energy band gap are explored.

1. INTRODUCTION:

According to the 1999 edition of the International Technology Roadmap for Semiconductors (ITRS)[1], continued scaling of CMOS devices beyond the 100nm technology node (2002) will rely on fundamental changes in transistor gate stack materials. Current research is being driven by the search for post-SiO_2 gate dielectrics. The replacement of SiO_2 with high-κ gate dielectrics (metal oxides, metal silicates, etc.) will also necessitate the use of gate electrode materials (other than the conventional dual-doped polysilicon) that form thermodynamically stable interfaces with the dielectric. Also driving the change to alternative gate electrodes is the growing need for tighter threshold voltage control and lower gate resistivity in short channel CMOS devices. The most promising alternative gate electrode materials are thus likely to be metals or their immediate derivatives (nitrides, silicides, etc.).

The flexibility afforded by the use of polysilicon as a gate electrode material is immense. The relative ease with which the work function of polysilicon can be selectively modified on a wafer precludes the need to *deposit/etch* two different gate electrode materials. It is evident that this flexibility will be lost in moving to metal gate electrodes (unless mid-gap metal electrodes are used). Dual metal gate CMOS technology has been successfully demonstrated recently [2];

however the processing sequence remains rather complicated. A simpler process flow, identical to the conventional CMOS process flow is desired.

One approach to incorporating metal gates while retaining the flexibility afforded by polysilicon is to use a metal whose work function can be *engineered* with relative ease. The selective conversion of a deposited metal film to its nitride is one such alternative. It is not trivial to pin the Fermi level of a metal at desired energy values. However, our recent experiments with sputter deposited molybdenum have provided encouraging results in this direction.

A high melting point, low coefficient of thermal expansion and low electrical resistivity make Mo an ideal material to withstand thermal processing budgets normally encountered in a CMOS fabrication process. Mo is also thermodynamically stable in contact with SiO_2 up to 1000°C [3] and owing to its high melting point is also expected to be stable on other promising high-κ gate dielectrics ($HfSi_xO_y$, $ZrSi_xO_y$, etc.).

From a transistor performance point of view, the single most important physical property that a post-poly Si gate electrode must possess however is its electron work function at the dielectric interface. There is considerable variation in the reported values of the Mo work function and values ranging from 4.2V-4.7V [4,5,6,7] have been reported. This leads one to believe that the effective work function of Mo is a function of deposition conditions, underlying dielectric and subsequent thermal processing.

In this paper, we report on investigations of the dependence of the Mo work function on post-deposition thermal processing. Our results indicate that the work function of Mo after annealing in the temperature range of 400-900°C is always larger than that of as deposited Mo. This increase in the work function occurs fairly independently of annealing ambient and temperature. Annealing in argon ambient produced results similar to annealing in forming gas ambient even after taking the effects of oxide fixed charge into account. On the other hand, implanting the deposited Mo films with N^{+14} ions before thermal processing is seen to restrict this increase in the work function. The post-anneal work function is observed to be quite stable and immune to subsequent thermal processing. The remainder of this paper is organized as follows. Section 2 outlines the experimental work performed. Section 3 discusses the results of the various experiments in light of the electrical and other physical analyses. Section 4 summarizes the results and presents conclusions.

2. EXPERIMENT:

Mo gate capacitors were fabricated on lightly doped p-silicon substrates. Thermally grown SiO_2 was used as the dielectric. In order to determine the magnitude of fixed charge at the oxide interface and its effect on the flat band voltage, multiple oxide thicknesses were obtained on a single wafer. A few die on each wafer were implanted with N^{+14} ions (dose: $5E15/cm^2$). After gate definition, wafers were annealed in Ar ambient. Annealing was performed at 400°C and 700°C for 15 minutes each. In order to emulate thermal budgets normally encountered in a CMOS process flow, all wafers received a high temperature rapid thermal anneal (900°C, 10s) and subsequent sintering in forming gas (10% H_2) at 400°C. Electrical characteristics were measured after each annealing step. The results are summarized in the following section.

3. RESULTS:

The Mo work function at the SiO_2 interface was extracted using values of the flat band voltage for capacitors with varying oxide thickness. In a MOS system the flat band voltage V_{fb} is related to the metal work function Φ_M by the following relationship:

$$V_{fb} = \Phi_{MS} - \frac{Q_f}{\varepsilon_{ox}} t_{ox}$$

where Q_f indicates the magnitude of fixed charge density in the oxide, t_{ox} is the oxide thickness and ε_{ox} is the permittivity of SiO_2. On a plot of V_{fb} vs. t_{ox}, the intercept corresponds to the value of Φ_{MS}, the metal-semiconductor work function difference while the slope of the linear relationship between V_{fb} and t_{ox} is a measure of the fixed charge Q_f in the oxide. The metal work function Φ_M on a p-type substrate is calculated using the following equation:

$$\Phi_M = \Phi_{MS} + \left(\chi_{Si} + \frac{E_g}{2q} + \varphi_B \right)$$

where χ_{Si} is the electron affinity of the Si substrate, E_g is the Si energy band gap and φ_B is a measure of the potential difference between the intrinsic and doped Si Fermi level. Work function values obtained in this way should be considered to be fair approximations after accounting for uncertainties in the determination of V_{fb} and the contributions of interface trap states.

Figure 1 shows the above relationship between V_{fb} and t_{ox} for capacitors with as deposited Mo gates and Mo implanted with N^{+14} ions ($5E15/cm^2$). The results are summarized in Table I. Figure 2 shows the C-V characteristics after the various annealing steps. It can be seen that the effective work function of as deposited Mo (without any implantation) is ~4.4V, while annealing at high temperature (700°C, 900°C) in argon ambient increases the work function by ~0.6-0.7V. A post RTA sintering anneal at 400°C slightly reduces the work function. Increase in the work function of metals at dielectric interfaces on post deposition annealing of MOS structures has been observed earlier. Matsuhashi and Nishikawa [4] observed that the work function of Mo deposited on Ta_2O_5 increases by 0.3V on annealing in the temperature range 400°C-800°C. Although a clear explanation for this phenomenon is absent in recent literature, it is hypothesized that metal work functions at dielectric interfaces are dependent on the metal film morphology at the interface. The morphology of metal films deposited on dielectrics is known to vary with deposition temperature, pressure and technique (physical or chemical) [8,9]. Deposition conditions will also influence subsequent evolution of the film morphology and the magnitude of residual stress in the film. Mo films in this study were deposited by sputter deposition on heated substrates (250°C). Figure 3 shows the results of x-ray diffraction analysis of Mo films after deposition and annealing at various temperatures. The presence of a well-defined Mo (110) peak is evidence of a columnar grain structure with (110) planes parallel to the substrate. The columnar grain morphology is also clearly observable in the cross-sectional SEM view of Figure 3. Mo is also highly prone to oxidation at elevated temperatures [10] and the possibility of oxidation bringing about an increase in the work function at the interface cannot be discounted. MoO_2 is known to be conductive and hence as long as the MoO_2 is formed without chemical reaction at the interface, oxidation is unlikely to alter the oxide capacitance. Dipole layers formed at metal-dielectric interfaces also contribute to the metal work function at the interface.

Local changes in crystalline orientation at the interface on thermal processing will thus affect the metal work function values through changes in the dipole moment at the interface.

Capacitors with N^{+14} implanted Mo gates did not show normal C-V characteristics up to 700°C annealing. This is most likely a result of extensive amorphization of the Mo films on exposure to a high implant dose. After annealing at 700°C however, these capacitors demonstrated normal C-V characteristics as shown in **Figure 2**. It can be observed that N^{+14} implantation induces a negative lateral shift in the C-V curves, indicating a reduction in work

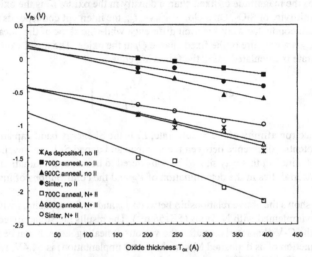

Figure 1. Determination of Φ_M before and after various thermal processing steps, filled symbols denote unimplanted samples while open symbols denote implanted ones.

function (~0.4V). In general, negative shifts in V_{fb} are indicative of increasing positive fixed charge in the dielectric. However, comparing the slopes of the V_{fb} vs. t_{ox} relationships for implanted and unimplanted capacitors does not reveal any significant change in the oxide fixed charge due to implantation. Implantation produces a near uniform reduction in V_{fb} regardless of the oxide thickness. It can thus be concluded that N^{+14} implantation into Mo reduces the effective work function. The same devices were then subjected to a 900°C RTA in Ar ambient to simulate thermal budgets during source/drain annealing in a conventional CMOS process flow. As seen in **Figure 2**, this high temperature anneal increases the flat band voltage by 0.4-0.5V, almost negating the effects of the prior 700°C anneal. Low temperature sintering does not further change the work function.

Table I. Mo work functions before and after thermal processing. Work functions of n+ and p+ polysilicon are ~4.0 V and ~5.1V respectively.

	No anneal	700°C	900°C	400°C Sinter
No II	4.42	5.1	5.05	5.0
Nitrogen II	-	4.03	4.41	4.42

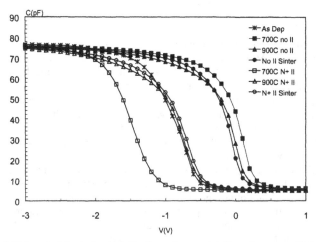

Figure 2. C-V characteristics before and after thermal processing. Capacitor area = 4E-4cm^2, t_{ox} = 185A.

A N^{+14} implant thus leads to a *metastable* low work function state in the MOS system. It is hypothesized that N^{+14} implantation changes the dipole charge distribution at the Mo/SiO$_2$ interface. Increase in the interface dipole charge density manifests itself in the C-V characteristics. This change however is not observed without sufficient "activation" of the implanted charge (brought about by annealing at 700°C). Concurrent with the activation of the implanted N^{+14} ions, the high temperature annealing is also likely to cause nitrogen outdiffusion, and the resulting dose loss can bring about an increase in the work function with an excessive thermal budget (900°C, 10s). Nevertheless, the implantation does have the effect of restricting the Mo work function to a value in the upper half of the Si energy band gap. Further optimization of the thermal processing and incorporation of suitable techniques to prevent nitrogen outdiffusion are expected to enable precise control of the Mo work function near that of n+ polysilicon. The dependence of the work function on implant dose and energy also needs further investigation.

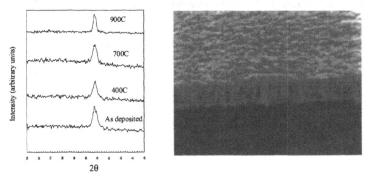

Figure 3. XRD analysis of Mo thin films deposited on SiO$_2$ before and after thermal processing and a cross-sectional SEM view depicting the Mo film morphology after 700°C annealing. The strong Mo(110) peak at 40.8° and the columnar microstructure are easily distinguishable.

4. CONCLUSIONS:

Gate electrode work function engineering is emerging as a critical requirement for continued scaling of deep sub-micron CMOS technology. Both bulk Si and SOI technologies present different gate work function requirements beyond the 100nm technology node. In order to minimize short channel effects, bulk devices need complementary gate work functions corresponding to E_c and E_v of Si for NMOS and PMOS devices respectively, while fully depleted SOI technology requires the gate work functions to be closer to the intrinsic Si Fermi level. The variation of the effective work function of Mo gate capacitors with thermal processing has been studied. The work function of Mo films increases on annealing at high temperatures (400C-900°C) and is fairly stable on post 900°C sintering. A local crystallographic rearrangement is believed to be responsible for this increase. Implantation of nitrogen is introduced as a novel technique to engineer the work function of Mo gate electrodes in CMOS devices. A medium dose implant of nitrogen (N^{+14}) into the Mo film before patterning produces a *metastable* low work function state. N^{+14} implantation followed by high temperature (700°C) annealing lowers the Mo work function to ~4.0V. Further high temperature annealing negates the earlier change to bring the Mo work function closer to the as deposited value. Further optimization of thermal processing and fabrication sequences needs to be performed to enable precise work function control.

REFERENCES:

1. International Technology Roadmap for Semiconductors, Semiconductor Industry Association, 1999
2. Q. Lu, Y. -C. Yeo, P. Ranade, T. -J. King, C. Hu, 2000 Symposium on VLSI Technology, IEEE, 2000, pp. 72
3. R. Beyers, Journal of Applied Physics, 56 (1), 1 July 1984, pp. 147
4. H. Matsuhashi and S. Nishikawa, Japan Journal of Applied Physics, v. 33, no. 3A, (1994), pp. 1293
5. R. G. Wilson, Journal of Applied Physics, v. 37, no. 8, July 1966, pp. 3170
6. P. Shah, IEEE Transactions on Electron Devices, v. ED-26, no. 4, April 1979, pp. 631
7. M.J. Kim and D.M. Brown, IEEE Transactions on Electron Devices, v. ED-30 no. 6, June 1983, pp. 598
8. T. Ishiguro and T. Sato, Materials Transactions, JIM, v. 39, no. 7 (1998), pp. 731
9. R. Brown, et al., IEEE Micro Electro Mechanical Systems Proceedings. An Investigation of Microstructures, Sensors, Actuators, Machines and Robots, Napa Valley, CA, USA, 11-14 Feb. 1990.), 1990. pp.77-81.
10. M.J. Kim and D.M. Brown, J. Electrochemical Society, v. 130, no. 10, 1983, pp. 2104

Mat. Res. Soc. Symp. Vol. 611 © 2000 Materials Research Society

STRUCTURAL AND CHEMICAL CHARACTERIZATION OF TUNGSTEN GATE STACK FOR 1 Gb DRAM

O. Gluschenkov[1], J. Benedict[1], L.A. Clevenger[1], P. DeHaven[1], C. Dziobkowski[1], J. Faltermeier[1], C. Lin[2], I. McStay[2], K. Wong[1]

[1]IBM Microelectronics, Semiconductor R&D Center, Hopewell Junction, NY;
[2]Infineon Technologies, Hopewell Junction, NY;
DRAM Development Alliance IBM/Infineon, IBM Semiconductor Research & Development Center, Hopewell Junction, NY.

Abstract

Material interaction during integration of tungsten gate stack for 1 Gb DRAM was investigated by Transition Electron Microscopy (TEM), X-ray Diffraction analysis (XRD) and Auger Electron Spectroscopy (AES). During selective side-wall oxidation tungsten gate conductor undergoes a structural transformation. The transformation results in the reduction of tungsten crystal lattice spacing, re-crystallization of tungsten and/or growth of grains. During a highly selective oxidation process, a relatively small but noticeable amount of oxygen was incorporated into the tungsten layer. The incorporation of oxygen is attributed to the formation of a stable WO_x ($x<2$) composite.

Introduction

Aggressive scaling of high density electronic memories, namely DRAM and eDRAM, calls for a low-resistance, low-aspect-ratio word lines. Under strict geometrical constraints, imposed by scaling, the resistivity of the line must be substantially reduced to maintain a high speed of the circuit. A multilayered word line having gate electrode, diffusion barrier, and gate conductor have been proposed in the literature.[1-5] A highly conductive gate conductor reduces the overall resistance of the structure while properties of the gate electrode are tailored to ensure the adequate transistor performance. The diffusion barrier prevents an interdiffusion of gate electrode and conductor material during high-temperature processing.

Due to its thermal stability, low diffusivity, and high conductivity tungsten is a leading candidate for use as a gate conductor material. There are several reports on the successful integration of W-based gate stack.[1-5] A typical W gate stack employs doped poly-crystalline Si as a gate electrode and WN_x as a sacrificial barrier. Thermodynamics of these materials predicts that during a high-temperature anneal (>800°C) the sacrificial barrier decomposes and reacts with Si to form a thermally stable diffusion barrier SiW_xN_y.[5,8] Upon word line patterning the gate stack needs a side-wall oxidation to repair the gate oxide damaged at the corners. The side-wall oxidation process is referred to as a selective side-wall oxidation process since it oxidizes Si selectively leaving the tungsten layer intact.[6-8] Although an acceptable performance of W-gate-stack transistor was demonstrated, interaction of materials during gate stack processing is not fully understood. This report is devoted to materials aspects of the W gate stack integration.

Experimental

The metal gate stack was fabricated on standard 8-inch (200mm) Si wafers. A standard trench-DRAM-compatible process flow was used up to the gate dielectric deposition. A thin SiO_2 layer was used as a gate dielectric. An in situ doped poly Si layer was deposited onto the SiO_2 layer by a CVD technique. Before metal/barrier deposition, the native oxide formed on the surface of poly Si was stripped in HF solution. WN_x sacrificial barrier was deposited by a PVD method using an Ar/N_2 ambient. Deposition of pure W was achieved by changing the PVD tool ambient to Ar gas only. The W layer was capped with a SiN film. After deposition, the gate stack was patterned and etched. Selective side-wall oxidation process was conducted in an Applied Materials Centura RTP system.[7,8] TEM cross section of the integrated W-stack is shown in Fig. 1.

Blanket monitor wafers were processed along with the patterned wafers for the analysis of stack materials modification and interaction during the processing. Processed blanket films were analyzed with X-ray diffraction (XRD), Auger electron spectroscopy, and resistivity measurement.

X-ray diffraction scans were run using monochromatic Fe radiation from a sealed tube source operating at 35kV and 35mA. Scans were run from 30 to 140° 2θ, with a step size of 0.1° and counting time/step of 20 sec. Auger depth profile analysis was performed on a Physical Electronics (PHI) Model 650 Scanning Auger Microprobe. The energy resolution was 0.25%. Beam voltage was 5 kV with a beam current of approximately 0.5 micro-amp. The beam was rastered over a square 100 square microns on the edge. The Auger depth profiles were accomplished with a differentially pumped argon ion sputter gun set for 3.5 kV. The sputter rate was set to remove 100A of thermally generated SiO_2 per minute. Data points were taken at intervals of 0.1 minute. Data reduction was done using the PHI MultiPak software package. Auger data was used in a 7 point differentiated format using peak to peak heights. Resistivity of tungsten films was calculated from the film sheet resistance. The sheet resistance was measured with a four point probe at 49 sites per wafer.

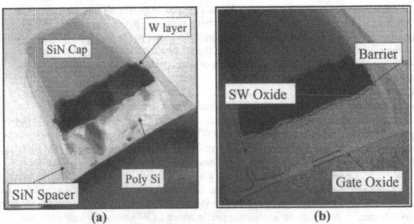

Figure 1. TEM cross sections of tungsten gate stack.

Results and Discussion

XRD analysis of uncapped W-stack was performed on W samples exposed to different annealing ambients. The anneal was used to simulate a thermal budget of the selective side-wall oxidation and was performed in a neutral ambient (N_2 or Ar) and in a typical ambient ($H_2:O_2$, 9:1, ~100 Torr) of selective side-wall oxidation process.

Fig. 2 compares XRD spectra of the three samples. The as-deposited W film shows a characteristic W pattern with a preferred orientation in the <110> planes. The W peaks in this spectrum are shifted to lower 2θ values as compared to a reference pattern for elemental tungsten. The diffraction peaks are relatively broad. Samples annealed in the neutral ambient show sharp tungsten diffraction peaks. Location of the diffraction peaks, match closely to the reference pattern for elemental W. Samples annealed in the selective oxidation ambient have a diffraction pattern similar to that of the samples processed in the neutral ambient.

The XRD analysis suggests that during a high-temperature anneal PVD W layer undergoes a structural transformation. As-deposited PVD W has a stretched crystal lattice and fine grain size. X-ray diffraction pattern of the tungsten is characterized by broad diffraction peaks shifted to lower 2θ. After high-temperature anneal the crystal lattice spacing becomes similar to that of the pure W and the size of W grains increases as evidenced by sharper tungsten diffraction peaks positioned at their expected location. The lattice spacing change suggests that Ar was incorporated into W lattice as a solid

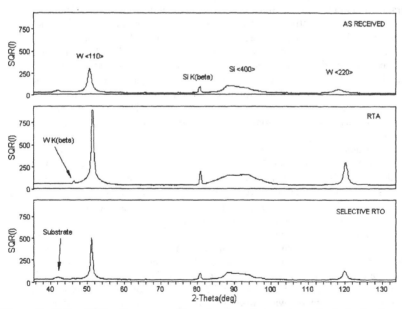

Figure 2. X-Ray Diffraction scans of tungsten layer in tungsten gate stack: (top) as-deposited, (middle) after neutral ambient RTA, (bottom) of selective RTO.

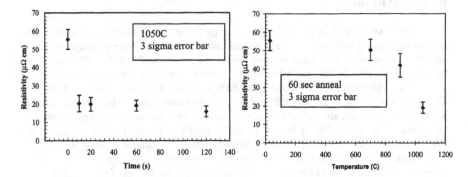

Figure 3. Resistivity of tungsten layer in tungsten gate stack as a function of annealing (left) time and (right) temperature.

state solution during the deposition and then was driven out during the anneal. The increase of W grains points to the presence of re-crystallization and/or grain growth during high-temperature anneals.

The structural transformation of W explains the observed dependence of W resistivity on anneals. Fig. 3 shows resistivity of W films as a function of anneal temperature and time. PVD W film annealed at a temperature higher than 1000°C for several seconds had a resistivity comparable to that of the pure tungsten. While the resistivity of as-deposited W films was 3 times higher than that of the pure tungsten.

Selectivity of the side-wall oxidation is demonstrated in Fig. 1b. A thick side-wall oxide is clearly visible on the poly-silicon wall while the exposed sides of the W-layer are intact. The side-wall oxide was grown at a high temperature (1050°C) and a low concentration of O_2 in the $H_2:O_2$ mixture (typically 5-20% of O_2).[8] It has been shown[7,8] that the selection of the process parameters, can make the oxidation process very selective, positioning it far from the border of selectivity at which the oxidation of W may occur.

Auger depth profiling was performed on W samples annealed in neutral and selective oxidation ambient. Fig. 4 shows Auger depth profiles for the samples annealed in the selective oxidation ambient (5% O_2) and pure N_2 ambient at 1050°C for 120 sec. In both cases there is no reaction between poly Si and the adjacent tungsten layer proving thermal stability of the diffusion barrier at 1050°C. In spite of its selectivity, the side-wall oxidation process introduces a noticeable amount of oxygen into the tungsten film. Although this amount of oxygen is not enough to produce the characteristic volume expansion of oxidized tungsten, it is clear that oxygen incorporates into the exposed tungsten during selective oxidation process. Such incorporation of oxygen into W layer occurs at the edges of the stack where W layer is exposed to the oxidizing ambient. Presence of oxygen in tungsten may reduce the gate conductor conductivity at the edges. Nevertheless, for 100 nm and wider word lines the increase of the stack resistance due this effect is small.

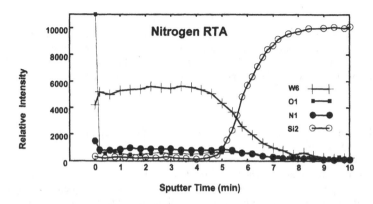

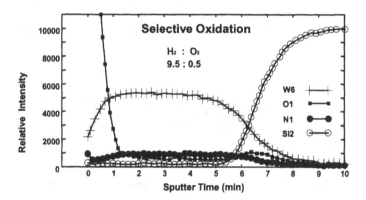

Figure 4. Auger depth profiles for tungsten gate stack annealed in (top) neutral ambient and (bottom) selective oxidation ambient.

One possible explanation to the oxygen incorporation into tungsten during selective oxidation process is a gradual nature of the boundary of selectivity. The boundary of selectivity is defined, based on the thermodynamic equilibrium of the following reaction[6-8]

$$WO_2 + 2H_2 = W + 2H_2O \tag{1}$$

With fixed free energies of reactants and products, the equilibrium is defined by the process temperature and the ratio of partial pressures of water vapor and hydrogen. Thermodynamics predicts a sharp boundary between oxidation of tungsten and reduction of WO_2. Fig. 4 shows that that the amount of oxygen in tungsten may vary gradually, that is, one may suggest an existence of WO_x ($x<2$) composite. This composite is, perhaps, a mixture of phases and/or oxygen atoms trapped on a tungsten lattice defect W:O. In any case since parameter x may vary the boundary of selectivity is defined with

$$WO_x + xH_2 = W + xH_2O \qquad (2)$$

rather than with Equation 1. In this case the boundary of selectivity may not be sharp in the temperature and partial pressure process space. Accordingly, process conditions which are highly selective in the sense of Equation 1 may allow for the formation of WO_x composite with parameter x determined by the processing conditions.

Conclusion

Low resistance tungsten gate stack was successfully integrated utilizing a sacrificial WN_x diffuison barrier and selective side-wall oxidation. XRD analysis revealed that tungsten gate conductor undergoes a structural transformation during selective side-wall oxidation. The transformation results in the reduction of tungsten crystal lattice spacing, recrystallization, growth of grains, and, consequently, a higher conductivity of the layer. Auger depth profiling showed that there is no silicidation of tungsten after selective side-wall oxidation suggesting that the diffusion barrier is stable at least at 1050°C. During highly selective oxidation process a relatively small but noticeable amount of oxygen was incorporated into the tungsten layer. The incorporation of oxygen cannot be explained in terms of the current view on the tungsten oxidation in mixtures of hydrogen and water vapor. Existence of a stable WO_x ($x<2$) composite may explain this effect.

References

1. M. Saito, N. Yamamoto, M. Yoshida, Y. Tanabe, T. Umezawa, H. Kawakami, T. Nagahama, N. Fukuda, Y.Hanaoka, K. Kawakita, T. Fukuda, T. Sekiguchi, Y. Tadaki, and N. Kobayashi, Materials Research Society, Conference Proceedings ULSI XIV, 625-630 (1999).
2. B.H. Lee, D. K. Sohn, J.S. Park, Y.J. Huh, S.B. Byun, and J.J. Kim, IEDM, 385 (1998).
3. Y. Hiura, A. Azuma, Y. Akasaka, K. Miyano, A. Honjo, K. Tsuchida, Y. Toyoshima, K. Suguro, and Y. Kohyama, IEDM, 393 (1998).
4. K. Ohishi, N. Yamamoto, Y. Uchino, Y. Janaoka, T. Tsuchiya, Y. Nonaka, Y. Tanabe, Y. Umezawa, N. Kukuda, S. Mitani, and T. Shiba, IEDM, 397 (1998)
5. K. Nakajima, Y. Akasaka, K. Miyano, M. Yakahashi, S. Suehiro, and K. Suguro, Appl. Surf. Sci., **117**, 312 (1997).
6. S. Iwata, N. Yamamoto, N. Kobayashi, T. Terada, and T. Mizutani, IEEE Trans. Electron Devices, **ED-31**, 1174 (1984).
7. B. Lin, M. Hwang, J.P. Lu, W.Y. Hsu, M. Pas, J. Piccirillo, G. Miner, K. O'Connor, G. Xing, D. Lopes, Mat. Res. Soc. Symp. Proc., vol. **525**, 359-364 (1998)
8. H. S. Joo, B. Ng, D. Lopes, G. Miner, Electrochemical Society Proceedings, vol. **99-18**, 203-209 (1999)

Advanced Gate Dielectrics

Mat. Res. Soc. Symp. Vol. 611 © 2000 Materials Research Society

A Closer "Look" at Modern Gate Oxides

Frieder H. Baumann, C.-P. Chang, John L. Grazul, Avid Kamgar, C. T. Liu, and David A. Muller
Bell Laboratories, Lucent Technologies
Murray Hill, NJ 07974

ABSTRACT

Using high resolution TEM (HRTEM), we identified some process induced 'weak spots' in SiO_2 layers: First, we observed thinning in the periphery of the transistor, i. e. near the boundary to the shallow trench isolation. At the boundary to the shallow trench, the Si substrate gradually changes its orientation from <100> to <110>, which results in an unexpected oxidation behavior in this region. Secondly, we observed the intrusion of poly-Si grains from the gate into the gate oxide, resulting in local thinning of the dielectric. Using image simulations, we show that conventional high resolution TEM can reveal the interface roughness only to a very limited extend.

INTRODUCTION

As the thickness of the gate oxide in high performance CMOS devices drops below 3 nm, local thinning on the atomic scale can have detrimental consequences on the reliability of the dielectric [1,2]. In addition, interfacial roughness at both sides of the dielectric constitute an increasing part the total oxide layer [3]. As an example, figure 1 shows the trend of SiO_2 gate oxides used in the personal computers owned by one of the

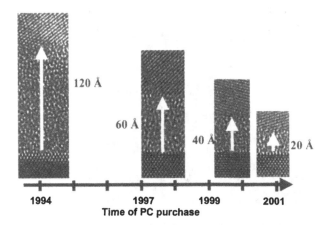

Figure 1. *Gate oxide thickness used in the personal computers owned by one of the authors (F. H. B.). All HRTEM images are taken from comparable in-house gate oxides.*

authors for the last five years. The figure shows the gate oxide thickness used in the central processing unit (CPU) vs. the time of purchase. (Although the 1 GHz computer has not been purchased yet, CPUs with 2 nm gate oxides entered the market in April 2000.)

The HRTEM images used in fig. 1 illustrate that gate oxides in modern integrated circuits have reached atomic dimensions, keeping in mind that 20 Å correspond to only 7 monolayers of silicon. Therefore, instrumentation is needed to measure and control the thickness of ultrathin films. Since the mid-1980s, HRTEM has been the instrument of choice for characterization, allowing thickness measurements on the atomic scale and in tiny volumes.

In this paper, we will demonstrate how HRTEM can be used to reveal severe thickness variations in modern Si processing. However, we will show that HRTEM can make only limited statements about the roughness and the roughness spectrum of the Si/SiO_2 interface. This will be accomplished by simulating images of samples with a known roughness spectrum.

EXPERIMENTAL DETAILS

All images shown in this paper have been obtained with a JEOL 4000 EX transmission electron microscope operating at 400 kV. Cross-sectional samples have been mechanically polished, and further thinned until perforation by Ar ion milling at 5 kV. Additional Ar ion polishing at 2.5 kV finished the specimen preparation.

EXPERIMENTAL RESULTS: Thickness variations

Figure 2 shows a cross-section of a transistor along its width. Here, the transistor gate leaves the active area to run over the shallow trench isolation (STI). In a shallow trench isolation process, smooth corner rounding has to be accomplished to avoid high

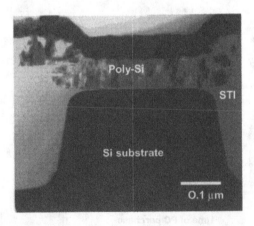

Figure 2. *Shallow trench isolation structure, showing ideal corner rounding*

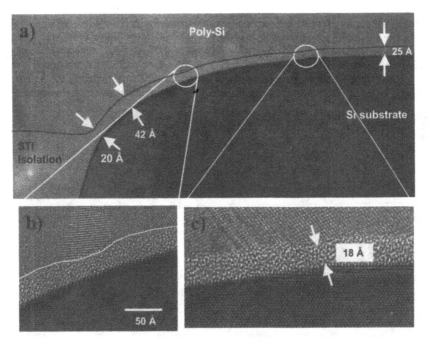

Figure 3. *a) High resolution TEM showing several thickness irregularities of the gate oxide near the shallow trench isolation. A nominally 25 Å thick gate oxide gradually increases as the orientation of the substrate deviates from <100>. b) Grains from the poly silicon gate intrude into the gate oxide. c) The corner rounding is accomplished by terracing, which can result in a thinning of the gate oxide down to 20 A.*

electric fields at sharp corners between silicon substrate and the gate material [4,5].

In figure 3, however, a similar cut is shown obtained from a wafer processed outside the correct processing window. Clearly, several areas in the gate oxide show severe non-uniformity. Fig. 3a) shows the gate oxide increasing from 25 to 42 Å, with a decrease down to 20 Å at the very corner to the STI. Fig. 3b) shows a poly-Si grain protruding into the gate oxide, causing local thinning. Fig. 3c) displays how the corner rounding at the STI corner is achieved: The Si substrate changes from the pure <100> orientation by building terraces, which can lead to a local thin spot in the gate oxide.

An even larger non-uniformity could be observed in a slightly thicker gate oxide. Figure 4 shows a thickening of the oxide from 37 to 57 Å as the orientation of the substrate changes from <100> in the channel to <111> at the corner. An increase in gate oxide thickness of 68% is expected due to the orientation dependence of oxide growth [13]. At the very corner, however, the gate oxide thickness is reduced dramatically down to 20 Å. A comparison of fig. 3 and fig. 4 shows thinning down to the value of 20 Å, both for a 25 and a 37 Å gate oxide. This seems to indicate that the thinning effect is independent of the oxidation time. The mechanism, which obviously restricts the oxide

C4.1.3

growth at the corner to the shallow trench, is so far unknown. A possible explanation for the observed thinning effect could be surface stresses, which build up during oxidation. Whether it is stress in the oxide hindering oxygen to diffuse to the silicon surface or stress in the substrate restricting further oxide growth cannot be decided at this point.

The possible electrical impact of severe oxide thinning is illustrated in fig. 5. Due to thinner oxide at the corner, the strips T2 and T3 of the main transistor T1 may have a

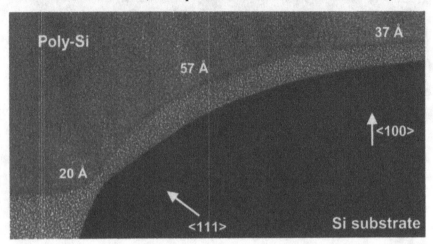

Figure 4. *Gate oxide non-uniformity near the shallow trench isolation corner. While the substrate orientation changes from <100> to <111>, the oxide thickness increases. At the very corner to the shallow trench, however, a dramatic thinning down to 20 Å is observed.*

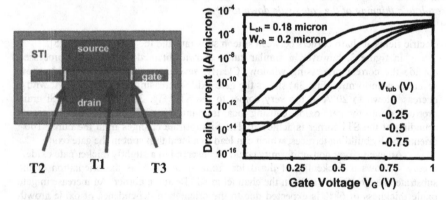

Figure 5. *Severe oxide thinning can result in different threshold voltages for certain parts of the transistor (T1 and T2, vs T3). That effect can lead to kinks in the subthreshold characteristics, as observed in the very small devices.*

lower threshold voltage, resulting in an earlier switch on and thus in a kink in the subthreshold characteristics. (For further details see ref. 6.)

EXPERIMENTAL RESULTS: Image simulations

HRTEM images are often used to determine the physical thickness of very thin layers, e. g. the physical thickness of a SiO_2 layer sandwiched between the silicon substrate and the poly-Si gate. In this section, we will investigate how well a high-resolution micrograph can define the position and the roughness of a crystalline-amorphous interface.

Basically, a HRTEM micrograph represents the projected potential of the part of the sample which the electron beam has passed through. Therefore, a crystalline part of the specimen will produce a lattice image revealing dots (atom columns) or lines (lattice planes), while the image of an amorphous specimen will consist of a speckle pattern [7]. Figure 6 shows schematically that even if the specimen is only partly crystalline, e. g. at a rough interface, the resulting lattice image will still reveal a periodic image.

Using image simulations, we investigated how the image of an interface with a known roughness spectrum changes when the sample thickness and the imaging conditions change. We used the NCEMSS imaging software package developed by Lawrence Berkeley Laboratory [8].

The virtual sample for the image simulations consisted of a slab of Si atoms, sandwiched between two SiO_2 layers. One interface was chosen to have a sinusoidal roughness of ~11 Å (5.43 Å excursions), and a roughness wavelength of 50 Å perpendicular, and 76 Å along the electron beam, respectively (see fig. 7).

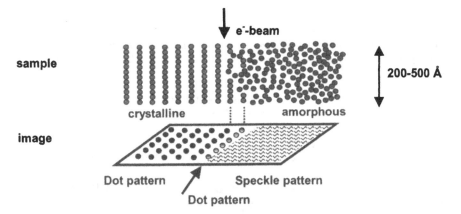

Figure 6. *The pattern formed in a HRTEM image represents the projected potential "seen" by the electrons passing through the sample. Therefore, crystalline parts of the sample produce periodic patterns, while amorphous parts produce a speckle pattern in the image. Even if the specimen is only partly crystalline along the path of the electron beam, still a periodic pattern can be produced.*

To perform the simulation of the imaging process, the position of each atom in our virtual sample has to be known. While this is trivial for the Si atoms in the Si slab (diamond structure), the atom positions for the silicon and oxygen atoms in the amorphous oxide are not readily known. We therefore took the following approach: Si atoms in the Si slab as represented as hard spheres. We then "deposited" differently sized hard spheres on both sides of the silicon slab, each deposited hard sphere representing a SiO_2 molecule (see fig. 8 and fig. 9). In order to achieve an amorphous layer and to prevent epitaxy, we used a well-known algorithm used to produce amorphous clusters with hard spheres [9]. In this algorithm, new spheres are sequentially deposited into cradles formed by three already existing spheres. The new sphere is always deposited into the cradle nearest to a predefined origin. This recipe leads to an amorphous cluster of hard spheres with a packing density of ~60%. The radius of the "SiO_2 sphere" was chosen to reflect the correct density of silicon oxide. In the image simulation program, the occupancy for a "SiO_2 sphere" was 33% silicon and 67% oxygen.

We believe that this rather simple approach in modeling the oxide gives valid

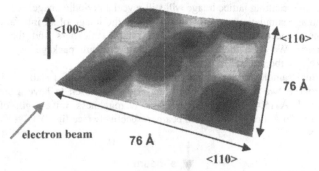

Figure 7. Interface roughness spectrum of virtual specimen used in image simulations. The Si surface has sinusoidal roughness with a wavelength of 76 Å in beam direction, and 51 Å in lateral direction.

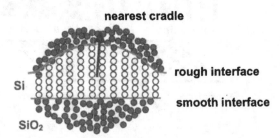

Figure 8. Principle of building amorphous layer around a crystalline layer: A new sphere (solid disk) is always deposited into the cradle (formed by existing spheres) nearest to a predefined origin. In this case, the origin is in the center of the Si slab (open symbols).

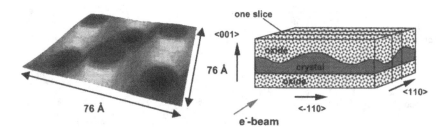

Figure 9. Final geometry of virtual specimen for image simulation. The sinusoidally rough Si surface builds the top part of a Si slab, which is sandwiched between two 50 Å thick SiO$_2$ layers. The lower interface is atomically sharp.

results for the matter under investigation, since only the projected potential of the oxide along the path of the beam is of interest. The immediate surrounding of each silicon or oxygen atom in the sample has little influence on the outcome of the simulation as soon as the sample thickness is larger than a few monolayers.

We performed image simulations for several sample thicknesses and imaging conditions. The virtual sample (see fig. 9) was cut into 15 Å slices, each slice containing roughly 5000 different atoms. The simulations were carried out on a SPARC 10 workstation. Imaging conditions were taken for a JEOL 4000EX TEM, operating at 400 kV with a spherical aberration constant of $c_s = 1$ mm. All images shown in this paper

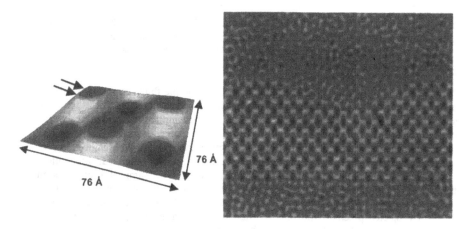

Figure 10. Image simulation of 15 Å thick cut through the part of the sample with the highest excursions (see arrows). Clearly, the simulated image shows a rough top interface, reproducing the roughness of about 11 Å peak to peak. The lower interface looks atomically sharp.

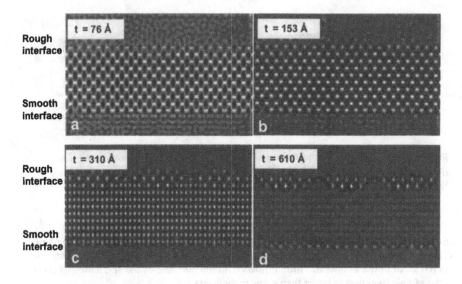

Figure 11. *Image simulations for various sample thicknesses. For a sample thickness equal to once (a) or twice (b) the roughness periodicity, the rough interface appears as smooth as the atomically rough bottom interface. For thicker samples, the rough interface reveals a differently pattern, and the lateral roughness spectrum seems to double (c)(d). In none of the above cases the true roughness is detected.*

have been simulated for an objective lens defocus of $\Delta f = -450$ Å, which is near the optimum (Scherzer) focus of the TEM used to obtain the experimental images.

Figure 10 shows a simulation of a 15 Å thick slice of the virtual specimen. Clearly, the surface roughness of the top surface ($+-5.4$ Å) is reproduced, while the bottom surface looks atomically smooth.

Figure 11a) shows an image simulation for a sample thickness of 76 Å, which corresponds to one roughness period in beam direction (see figure 9), while fig. 11b) shows a simulation for twice the sample thickness. Surprisingly, the top interface with a known roughness of 11 Å peak to peak appears in the image as smooth as the bottom interface, which is atomically abrupt by construction. Apparently, the imaging process smoothes the interface out, letting the top interface look as smooth as the bottom interface. Figure 11c) and d) show simulations for even thicker samples. It can be seen that the rough interface can lead to a pattern change. At the interface, the electron beam samples over less crystalline material than in the bulk of the Si slab, which results in a different projected potential and thus a different image. In addition, the images of the thicker samples reveal much smaller roughness amplitudes (~3 Å peak to peak), and a different roughness spectrum. It should be noted that the apparent location of the interface moves up into the oxide by a few Å.

Similar findings have been reported using image simulations with a different interface model and smaller roughness wavelengths [10,11]. Attempts to analyze the averaging process in more detail have been performed by Goodnick et al. [12].

The image simulations presented in fig. 10 and 11 clearly show that only for samples, which are much thinner than the dominant roughness periodicity, the true roughness is revealed. However, practical TEM samples tend to be thicker than 100 Å, in order to avoid artifacts from specimen preparation (ion milling damage). Therefore, extreme caution has to be used when the roughness of the interface between crystalline and amorphous materials is judged by means of HRTEM images.

It should be noted that HRTEM is able to determine the roughness of interfaces between crystalline materials. The method of Chemical Mapping [14] uses known intensity changes in the image to quantify how the chemical content of an atom column changes, and thus is able to make reliable accounts of the interface roughness.

Recently, Scanning Transmission Microscope (STEM) images have shown that interface roughness in an α-c system can be measured quantitatively [3]. Muller et al. could show that by using annular dark field images and electron energy loss spectroscopy (EELS), a more sophisticated and quantitative analysis of the Si/SiO$_2$ interface is possible.

EXPERIMENTAL RESULTS: Poly-intrusion into the gate oxide

Traditionally, the poly-Si/SiO$_2$ interface has been considered as less important than the Si-substrate/SiO$_2$ interface, mainly because the channel for the charge carriers is located near the latter interface. But by entering the regime of only a few monolayers of

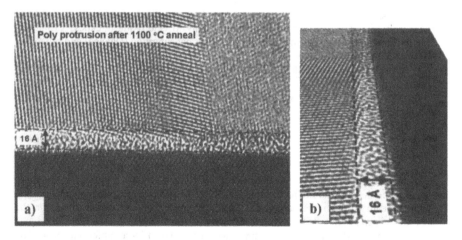

Figure 12. a) *High resolution TEM image of 16 Å gate oxide. Lattice plains from a poly-Si grain intrude into the gate oxide. b) Slanted reproduction of a), showing that the poly-Si lattice plane intrude up to 8 Å into the gate oxide.*

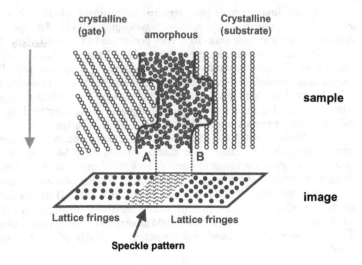

crystalline (gate) amorphous Crystalline (substrate)

sample

A B

image

Lattice fringes Lattice fringes

Speckle pattern

Figure 13. *Alternative explanation for poly-Si lattice planes intruding into the gate oxide. Shown is a hypothetical dent in the Si substrate, covered by a conformal oxide and poly-Si. Although the oxide has the same thickness everywhere, the imaging process (projected potential) may depict a locally thinner gate oxide in the micrograph (for details, see text).*

dielectric between the gate and the channel, roughness or other physical non-uniformities at the poly-Si interface can contribute substantially to gate oxide thickness variations. That in turn can have severe implications on power consumption (leakage) and reliability (dielectric breakdown).

We therefore had a closer look at the poly-Si interface in a process, where the dopant activation anneal was executed at a relatively high temperature of 1100 °C. Figure 12a shows a HRTEM image of a 16 Å gate oxide, where lattice plains from the poly-Si seems to intrude into the channel by about 5-8 Å. (The two dotted parallel lines in the image are guides for the eye.) This intrusion of lattice plains into the gate oxide is even more obvious in fig. 12 b, which shows a slanted reproduction of the same micrograph.

Although the intrusion of the poly-Si into the gate oxide depicted in figure 12 is very obvious, the HRTEM image alone is not a sufficient proof that intrusions like these exsist. Since the electron microscope images the projected potential along the path of the electron beam through the sample, at least one other possible explanations for the observation can not be ruled out completely [15]:

Imagine a dent in the Si-substrate, which is covered by a conformal oxide. The oxide itself is covered by poly-Si (see figure 13). Since in HRTEM the projected potential is imaged, both interfaces A and B of the dielectric will show lattice fringes due to the fact that part of the material that the beam passed through is crystalline. Therefore, this effect may show a local thinning in the oxide, although the oxide just follows the roughness of the substrate and has the same thickness everywhere.

In the case of fig. 12, possible misinterpretations of a HRTEM image can be revealed by high resolution EELS mapping of the interface (for details, see [3]): If no oxygen is detected in region B (interface on substrate side), then the substrate is flat, the oxide is in deed thinner locally, and the image of area A represents an intrusion.

CONCLUSIONS

Very thin gate oxides can reveal severe trouble spots when the substrate deviates from the ideal <100> orientation. One example is the corner near the shallow trench isolation. Here, we observed local thinning of the gate oxide up to 50%, which can have detrimental effects on transistor leakage and reliability.

Imaging the roughness of amorphous-crystalline interfaces with conventional high-resolution TEM methods turns out to be a difficult task. Image simulations of a Si/SiO_2 interface with a known and simple roughness spectrum show that quantification of the roughness is not possible by simple visual inspection of the micrograph. It turns out that the imaging process seems to smooth out the interface roughness, at least along the beam direction, resulting in much less apparent roughness than initially defined.

Finally, in several occasions intrusions of the poly-Si into the gate oxide have been observed in our lattice images. Although corroborating evidence from EELS mapping is still missing, these possible intrusions (up to 8 Å for a 16 Å gate oxide) can have serious consequences for reliability and leakage of modern CMOS devices, and for the ultimate limit of gate oxide shrinking.

ACKNOWLEDGEMENTS

The authors would like to thank P. O'Sullivan for help with the graphics in the manuscript, and D. J. Eaglesham for helpful discussions.

REFERENCES

1. B. E. Weir et al., *1997 IEDM Technical Digest*, **73**, (1997)
2. G. L. Timp et al., *1997 IEDM Technical Digest*, **930**, (1997)
3. D. A. Muller, T. Sorsch, S. Moccio, F. H. Baumann, K. Evans-Lutterodt, and G. Timp, *Nature*, Vol. 399, 6738, **758**, (1999)
4. M. R. Pinto, W. M. Coughran, C. S. Rafferty, E. Sangiorgi, and R. K. Smith, In: Hess K, Leburton J P, Ravaioli U (eds.): *Computational Electronics*, Kluwer Academic Publishers, Norwell, MA. (1991)
5. C. P. Chang et al., *1997 IEDM Technical Digest*, **661**, (1997)
6. C. T. Liu, F. H. Baumann, A. Ghetti, H. H. Vuong, C. P. Chang, K. P. Cheung, J. I. Colonell, W. Y. C. Lai, E. J. Lloyd, J. F. Miner, C. S. Pai, H. Vaidya, R. Liu, J. T. Clemens, *1999 VLSI Symposium*, (1999)
7. J. C. H. Spence, *Experimental High-Resolution Electron Microscopy*, 2. Edition, Oxford University Press, New York Oxford, (1999)

8. M. A. O'Keefe and R. Kilaas, *Users Guide to NCEMSS,* and references therein, http://ncem.lbl.gov/frames/software.htm
9. C. H. Bennett, *J. Appl. Phys.,* Vol. 43, No. 6, **2727**, (1971)
10. Ohdomari, T. Mihara, and K. Kai, *J. Appl. Phys.* 60 (11), **3900**, (1996)
11. H. Akatsu and I. Ohdomari, *Appl. Surf. Science* 41/42, **357**, (1989)
12. S. M. Goodnick et al., *Phys. Rev. B,* Vol. 32, No. 12, **8171**, (1985)
13. S. Wolf and R. N. Tauber, *Silicon Processing for the VLSI Era, Vol. 1,* Lattice Press, Sunset Beach, California, **212**, (1986)
14. A. Ourmazd, F. H. Baumann, M. Bode and Y. Kim, *Ultramicroscopy,* Vol. 34, 237, (1990)
15. H. Feichtinger, *"The devil never sleeps. ",* private communication, (1987)

Mat. Res. Soc. Symp. Proc. Vol. 611 © 2000 Materials Research Society

Low-Oxygen Nitride Layers Produced by UHV Ammonia Nitridation of Silicon

Mark A. Shriver[1], T.K. Higman[1], S.A. Campbell[1], Charles J. Taylor[2], and Jeffrey Roberts[2]
[1]Department of Electrical and Computer Engineering, 200 Union St SE,
[2]Department of Chemistry, 207 Pleasant Street SE,
University of Minnesota,
Minneapolis, Minnesota 55455, U.S.A.

ABSTRACT

If chemically vapor deposited high permittivity materials such as TiO_2 and Ta_2O_5 are to gain wide acceptance as alternatives to SiO_2 gates in silicon MOSFETs, the interface between the deposited high-k material and the silicon must be abrupt and have a low density of electrically active defects. Unfortunately, the process for depositing these materials often produces an unacceptably thick, low-permittivity amorphous layer at the interface, which reduces the effectiveness of the high-k material and often contains unacceptably large numbers of charge states. One way to prevent this layer from forming is to deliberately introduce a very thin layer of Si_3N_4 to act as a diffusion barrier prior to deposition of the high-k material. Previous work has shown nitrides to have high concentrations of traps and interface states, but these films also had considerable oxygen contamination, particularly at the nitride-silicon interface. In this paper, we show that direct thermal nitridation of the silicon surface in ammonia can provide a low interface state density surface that is also an excellent diffusion barrier. A key feature of this process is the various techniques needed to obtain very low oxygen incorporation in the Si_3N_4. Even at the Si_3N_4-Si interface, the oxygen content was near the detection limits (0.5%) of Auger Electron Spectroscopy (AES). The nitride films were grown in a range of temperatures that resulted in self-limited thicknesses from a few monolayers to a few nanometers. These films were then characterized by Auger, Time-of-Flight SIMS, and in the case of the thicker films, capacitance-voltage techniques on both n- and p-type silicon substrates. The data shows very low levels of oxygen contamination in the nitride films and low interface state densities in capacitors fabricated from this material.

INTRODUCTION

Silicon dioxide, the dominant dielectric material of silicon based MOSFET devices, will not withstand scaling to the degree needed for future gate materials. Smaller scaling is needed to increase the speed at which MOSFET's operate. The next generation of MOSFET devices will prosper using high-permittivity (high-k) materials such as ZrO_2, TiO_2, and HfO_2. These films can produce a gate-stack capacitance that is much higher than the SiO_2 stacks currently used in production. Although high-k oxides can give a higher capacitance per unit thickness than SiO_2, these materials will form SiO_2 during growth at the high-k-Si substrate interface [1]. This interfacial SiO_2 layer will degrade the performance of the gate and will continue to grow during high temperature processing. The poor interface of high-k materials grown on Si substrates has defects sites, which will trap charge shifting the threshold voltage and degrading the mobility in the channel.

In order to prevent the formation of SiO_2 at the gate insulator-Si substrate interface, a thin diffusion barrier is needed. This thin film must furthermore have low defect density to prevent charge trapping near the channel. Properly grown silicon nitride (Si_3N_4) films have both good diffusion barrier properties and a low interface state density. Although the interface state density of SiO_2 is quite good, interface densities as low as $10^{10}/cm^2/eV$ on (100) Si have been observed with UHV nitrides [2]. It should also be noted that Si_3N_4 has lower threshold shift under bias stress compared to SiO_2 [3]. The formation and characteristics of thermal nitrides are well understood [4-8] and the electrical properties of silicon nitride can be improved by minimizing the oxygen content thereby preventing defect states and the formation of SiO_2 at the Si_3N_4-Si interface.

Silicon nitride is a promising material for interfacial layers not only because of extensive previous investigations into the material, but also because it can be directly substitutable for oxides in many production processes. Silicon nitride is self-limiting in growth with temperature determining the ultimate thickness. This characteristic leads to a consistent and easily reproducible process. The first few monolayers of Si_3N_4 form very quickly, then growth decreases because nitrogen must diffuse farther through the growing film to react with the silicon [7,8]. This self-limited growth could make it possible to grow a thin interfacial layer quickly, consistently, and at a moderate temperature, which are properties desirable for an IC fabrication process.

EXPERIMENTAL DETAILS

In this paper, nitride films were grown using thermal nitridation of silicon in ammonia gas under UHV conditions. Using this method there are three possible sources of oxygen incorporation into the film: the Si substrate, the ammonia, and the ambient gas in the chamber. In order to reduce possible oxygen contamination from the Si, only float-zone Si was used. Possible oxygen contamination from the ammonia gas was eliminated by using an Aeronex Gatekeeper hydride gas purifier.

The UHV system used had a base pressure of 10^{-9} Torr and was attached to an analytical chamber which allowed *in situ* Auger electron spectography to be performed. Oxygen was reduced by baking the chamber at 180 C for 72 hours or more with purified ammonia flowing into the chamber. Then a titanium sublimation pump was flashed for 1 min to deposit a fresh layer of gettering material. The system was then baked for several more hours before nearly oxygen-free films were made. The correct bake-out procedures were crucial to producing low-oxygen films.

The wafers were 4-inch (100) Si p- and n-type wafers with 14-20 and .087-.113 Ω cm resistivity, respectively, grown using the float-zone method. In order to improve contact on the backside of the wafers, P-type wafers were backside implanted with boron and n-type wafers with phosphorus both with a dose of $5x10^{15}$ cm^{-2} and an energy of 60 keV. The wafers had to be cleaved into 3 cm squares in order to fit into the UHV system. Before entering the chamber, the samples were soaked in piranha solution (7:3 H_2SO_4:H_2O_2), rinsed with distilled water, blown dry with compressed air, dipped in a solution of 10:1 H_2O:HF and blown dry without a rinse. This prepared a hydrogen-terminated surface on the surface of the wafer.

The silicon samples were then pumped down to 10^{-7} Torr in a load lock before being transferred into the deposition chamber. The samples were heated to 900 C as measured by pyrometry under vacuum before being nitrided at a pressure of 10^{-5} Torr for 5 min. The samples were then cooled in ammonia. Auger analysis was then performed *in situ* to determine the relative oxygen concentration in the film. The auger electron beam energy was set high enough to completely penetrate the film.

Metal-Insulator-Semiconductor (MIS) capacitors were fabricated using doped polysilicon as the metal layer. The poly-Si dopants were activated by a 30 second rapid thermal anneal (RTA) at 900 C. The capacitance and leakage current were then measured to determine the quality of the films.

DISCUSSION

Auger electron spectroscopy (AES) analysis, figure 1, showed that the oxygen content in the films was as low as about 0.5%, which is near the detection limits of Auger. This gives a relative atomic nitrogen concentration ([N]/[N]+[O]) of 98%, which is higher than the value of 90% reported in a recent study done in a high vacuum (base pressure 10^{-5} Torr) system [3]. The Auger film composition results do not seem to indicate stoichiometric Si_3N_4 because the Auger e-beam penetrates beyond thickness of the film into the Si substrate.

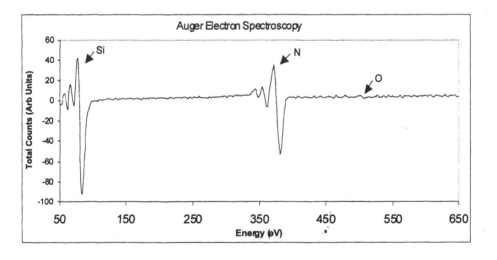

Figure 1. *Auger Electron Spectroscopy(AES) analysis of Si_3N_4 film grown at 900 C for 5 min. Although an oxygen peak was detected, it was less than 1% of the overall content of the film.*

The films grown in this study were 1.7 - 2.5 nm thick measured by ellipsometry with the refractive index fixed at 2.0. This agrees with the thickness determined from cross-sectional TEM and AFM imaging of similarly grown UHV nitride films [13].

The capacitance vs. gate bias plot shown in figure 2 was produced using a UHV nitride MIS capacitor. Due to gate leakage through the very thin Si_3N_4, the overall shape of the C-V curve is distorted and extraction of interface state and trapped charge densities by detailed comparison to the theoretical C-V curve is inappropriate. The plot is included here for qualitative comparison only. While Auger analysis of this sample showed some carbon and oxygen contamination which would tend to induce a slight threshold shift and some interface states, these values are low based on the qualitative analysis of the C-V trace.

Both p- and n-type Si substrates produced similar quality films. Threshold voltage shifts due to hysteresis below 5 mV have been observed on our UHV nitride samples.

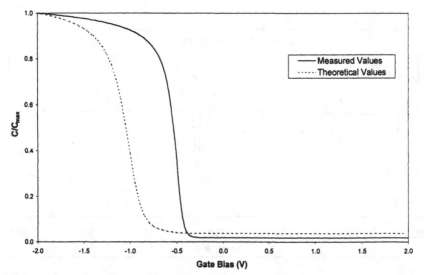

Figure 2. *Capacitance vs. gate bias plot for a MIS capacitor formed with a UHV nitride insulator. The plot was produced at 500 kHz.*

Time-of-flight Secondary Ion Mass Spectroscopy (TOF SIMS) data, figure 3, displays oxygen on the surface of the nitride which may have come from a native oxide layer which formed when the sample was removed from the chamber. Little oxygen is seen at the Si_3N_4-Si substrate interface.

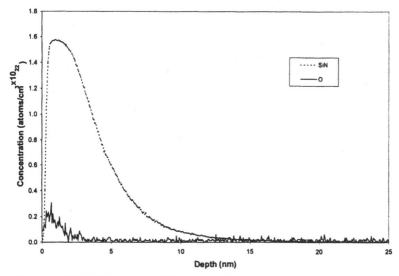

Figure 3. TOF SIMS data for UHV nitride. The oxygen in the film is near the surface and is mostly due to a native oxide formed during atmospheric exposure.

CONCLUSIONS

We have studied the compositional and electrical properties of silicon nitride films by AES, TOF SIMS, ellipsometry, and C-V measurements. While more study is needed to accurately determine interface state and trapped charge densities, these results are promising. We have shown that Si_3N_4 films with low levels of oxygen and carbon contamination can be produced which also exhibit low interface state densities. Ultimately, complete gate stacks comprised of high-k material with a UHV nitride interfacial layer formed on Si will have to be formed for a complete analysis.

ACKNOWLEDGEMENTS

The authors would like to thank Dr. Chris Hobbs at Motorola for the TOF SIMS data. Research was supported by the Semiconductor Research Corporation under research ID number 616.018.

REFERENCES

1. S. A. Campbell, B. He, R. Smith, T. Ma, N. Hoilien, C. Taylor, and W. L. Gladfelter, Group IVB Oxides as High Permittivity Gate Insulators, *Proc. MRS Fall 1999 Meeting* (1999).
2. T. K. Higman, R. T. Fayfield, M. S. Hagedorn, K. H. Lee, S. A. Campbell, J. N. Baillargeon, and K. Y. Cheng, *J. Vac. Sci. Technol. B,* **11**, 992 (1993).
3. H. Shin, S. Shoi, T. Hwang, and K. Lee, *J.Korean Phys. Soc.*, **33**, S175 (1998).
4. M. Moslehi, and K. Saraswat, *IEEE Transactions on Electron Devices.*, **32**, 106 (1985).
5. S. P. Murarka, C. C. Chang, A. and C. Adams, *J. Electrochem. Soc.*, **126**. 996 (1979).
6. Y. Hayafuji, and K. Kajiwara, *J. Electrochem. Soc.*, **129**, 2102 (1982).
7. I. J. R. Baumvol, F. C. Stedile, J.-J. Ganem, S. Rigo, and I. Trimaille, *J. Electrochem. Soc.*, **142**, 1205 (1995).
8. I. J. R. Baumvol, L. Borucki, J. Chaumont, J.–J. Ganme, O. Kaytasov, N. Piel, S. Rigo, W. H. Schulte, F.C. Stedile, I. Trimaille, *Nucl. Intr. and Meth. in Phys. Res. B,* **118**, 449 (1996).
9. R. T. Fayfield, J. Chen, M. S. Hagedorn, T. K. Higman, A. M Moy, and K. Y. Cheng, *J. Vac. Sci. Technol. B* **13**, 786 (1995).

Mat. Res. Soc. Symp. Vol. 611 © 2000 Materials Research Society

LOW TEMPERATURE NITRIDATIO OF SiO$_2$ FILMS USING A CATALYTIC-CVD SYSTEM

Akira Izumi, Hidekazu Sato and Hideki Matsumura
JAIST (Japan Advanced Institute of Science and Technology)
Ishikawa 923-1292, JAPAN

ABSTRACT

This paper reports a procedure for low-temperature nitridation of silicon dioxide (SiO$_2$) surfaces using species produced by catalytic decomposition of NH$_3$ on heated tungsten in catalytic chemical vapor deposition (Cat-CVD) system. The surface of SiO$_2$/Si(100) was nitrided at temperatures as low as 200°C. X-ray photoelectron spectroscopy measurements revealed that incorporated N atoms are bound to Si atoms and O atoms and located top-surface of SiO$_2$.

INTRODUCTION

Surface nitridation of SiO$_2$ is useful for metal-oxide-semiconductor (MOS) gate dielectric to suppress boron penetration from p$^+$-Si gate [1]. Such treatment should be prepared at low temperatures without plasma enhancement to suppress dopant diffusion and plasma damage. Excessive nitrogen at the SiO$_2$/Si interface may reduce peak carrier mobility in the channel of MOS devices [2]. Therefore, top-surface nitridation of SiO$_2$ is required for the suppression of boron penetration without degradation of device.

In this paper, we propose a low temperature and plasma-less top-surface SiO$_2$ nitridation technology using species produced by catalytic decomposition of NH$_3$ on heated W in catalytic chemical vapor deposition (Cat-CVD) system.

In the Cat-CVD method, deposition gases are decomposed by catalytic cracking reactions with heated catalyzer placed near substrates so that films are deposited at low substrate temperatures without help from plasma excitation. The authors have succeeded in depositing high quality SiN$_x$ films using a SiH$_4$ and NH$_3$ gas mixture [3]. High quality amorphous Si [4] and polycrystalline Si films [5] with low hydrogen content have also been obtained using a SiH$_4$ and H$_2$ gas mixture. Applying this technique, we have succeeded in direct surface nitridation of Si at 200°C using NH$_3$ gas [6]. This method is also useful for surface cleaning [7] and nitridation of GaAs [8]. Crystalline Si etching with high etch rate as high as 200nm/min is obtained using H$_2$ gas [9]. This present work is on a similar way of such wide variation of the Cat-CVD application.

EXPERIMENTAL

Figure 1 shows the schematic diagram of the apparatus for nitridation. The apparatus is

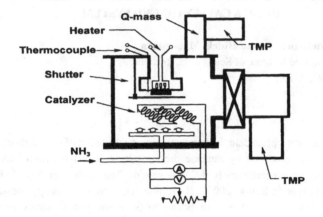

Fig.1 Schematic diagram of the apparatus for nitridation.

basically same as the Cat-CVD apparatus except for the source gases. A tungsten (W) wire with a diameter of 0.5 mm and length of 1300 mm was used as the catalyzer. It was coiled and spread widely in a space of 70 mm x 70 mm, keeping it parallel to a substrate holder with a distance of 60 mm. Therefore, to make the spread area of catalyzer larger, it is possible to make nitridation on large diameter wafers.

An n-type CZ Si(100) wafer with a resistivity of 0.85-3.0 Ωcm was cut into 2x2 cm^2 pieces. After that they were degreased and cleaned by RCA method. Then, they were dipped in 2% diluted HF for 2 min. After the cleaning, the pieces of wafer were oxidized by dry O_2 at 900°C and thin SiO_2 films (4.2 nm) were formed on them. Then, some pieces of the $SiO_2/Si(100)$ were immediately loaded into the chamber. After that the chamber was evacuated to 5×10^{-7}Torr. An NH_3 gas with 99.999% purity was used for surface nitridation of $SiO_2/Si(100)$. The gas pressure during nitridation was about 0.1 Pa. The flow rate of NH_3 was fixed at 50 sccm. The temperature of the substrate holder (T_s), which was measured by a thermocouple (TC) mounted on the same surface of the holder and just beside the substrate, was always kept at about 150°C. The surface temperature was calibrated using a TC directly mounted to the Si substrate, and it was confirmed that the surface temperature was at most 50°C higher than T_s. The temperature of W catalyzer (T_{cat}) which was monitored by infrared thermo-meter was kept at 1800°C. The surface conditions of $SiO_2/Si(100)$ were characterized by *ex-situ* x-ray photoelectron spectroscopy (XPS) measurements using monochromatic Al K_α radiation. Most of spectra were observed at the photoelectron take-off angle θ of 35°. The shift of signals due to electrical charging of the sample surface was corrected with the C(1s) signal from the carbon contaminant. The samples were characterized by capacitance versus voltage (C-V) measurements. Metal-insulator-semiconductor (MIS) diode structures for C-V

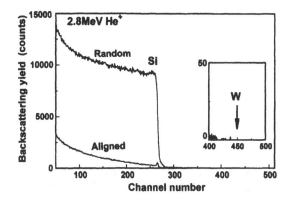

Fig.2 RBS spectra of Si(100) after 30 min. nitridaton treatment.

measurements were completed by evaporating Al on $SiO_2/Si(100)$ through a metal mask. In these structures, no post metal annealing treatment was performed. The electrode area was $3\times10^{-4}cm^2$. C-V measurements were performed at 1 MHz in the dark at room temperature. The bias voltage was scanned from the depletion-inversion side to the accumulation side at a sweep rate of 0.1 V/s.

RESULTS AND DISCUSSION

In this method, the W catalyzer decomposes the NH_3 gas. Therefore, contamination of the surface of SiO_2 with W might be considered. Figure 2 shows Rutherford backscattering

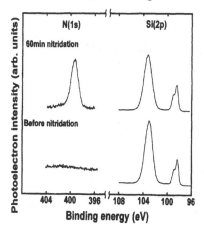

Fig.3 XPS spectra of Si(2p) from the surface of Si(100), (a): before oxidation and (b): after 60 min. oxidation.

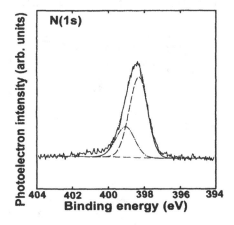

Fig.4 Decomposed XPS spectrum of N(1s) from the surface of Si(100) after 60 min. nitridation.

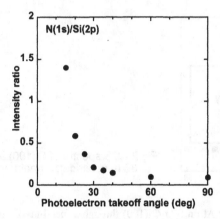

Fig.5 Intensity ratio of N(1s)/Si(2p) against photoelectron take-off angle.

(RBS) spectra of Si(100) for 60 min. nitridation. RBS measurements were performed using 2.8 MeV He ions. W atoms cannot be detected at the position of W, channel number of 450, in the spectrum. It is evaluated that the atomic ratio of the W atoms to Si is at least a few ppm or less by taking into account of the cross section of Si and W to He, even if they do exist. This implies that the W contamination is negligible.

Figure 3 shows XPS spectra of N(1s) and Si(2p) from the surface of Si(100), (a): before nitridation and (b): after nitridation for 60 min. As can be seen, the peak of N(1s) spectrum appears after the nitridation treatment. The N(1s) XPS spectrum can be decomposed into two major components as shown in Fig.4. One of the components at 398.5eV is derived from -N-Si$_2$ bonds [10] and another is derived from O-N-Si$_2$ [11].

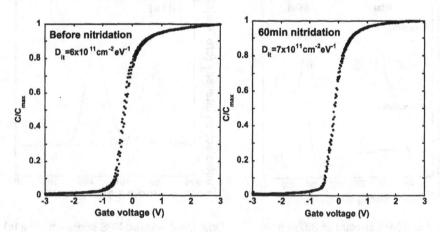

Fig.6 C-V curves of before and after 60min. nitridation

Above result indicates that the Incorporated nitrogen atoms are bound to silicon and oxygen in the SiO_2 films. According to the aerial intensities of these XPS spectra, we can estimate that the SiO_2 film is nitrided about 10 at. %.

Angle-resolved XPS (ARXPS) measurements were performed as shown in Fig.5 to determine the position of nitrided layer with varying the photoelectron take-off angle θ, which was defined as the angle between the XPS detector and the sample surface, in the range between 90 and 15 degree. Intensity ratio of N(1s)/Si(2p) increases as θ decreases. Smaller θ gives information in the region nearer to the surface. Therefore, the position of the nitrided layer is located at the top-surface of SiO_2.

Figure 6 shows the C-V curve measured on the MIS diodes of SiO_2/Si(100) before and after nitridation for 60 min. Before the nitridation treatment, the C-V curve shows a small injection-type hesterisis. However, after the nitridation , the hysterisis of the C-V was almost disappeared. The minimum values of the interface state density calculated by the Terman method [12] before and after the nitridation treatment are $6x10^{11}$ $cm^{-2}eV^{-1}$ and $7x10^{11}$ $cm^{-2}eV^{-1}$, respectively. It is remarkable that the flat-band voltage shift was not observed even after the nitridation treatment.

CONCLUSIONS

A new procedure for low-temperature nitridation of SiO_2 surfaces using NH_3 through a W catalyzer has been developed. Following results are obtained.

1) The surface of SiO_2 is nitrided at temperatures as low as $200^{\circ}C$.
2) The incorporated N atoms are bound to Si atoms and O atoms and located top-surface of SiO_2.
3) The flat-band voltage shift of C-V characteristics was not observed after the nitridaion treatment.

ACKNOWLEDGMENT

The authors would like to express their thanks to Prof. S. Horita and Dr. A. Masuda at JAIST for their fruitful discussions. This work is in part supported by the R&D Projects in Cooperation with Academic Institutions "Cat-CVD Fabrication Processes for Semiconductor Devices" entrusted from the New Energy and Industrial Technology Development Organization (NEDO) to the Ishikawa Sunrised Industries Creation Organization (ISICO) and carried out at Japan Advanced Institute of Science and Technology (JAIST). This work was also in part supported by Ozawa and Yoshikawa Memorial Electronics Research Foundation, the Foundation of Ando Laboratory and Grant-in-Aid for Scientific Research from the Ministry of Education, Science, Sports and Culture.

REFERENCES

1. E. P. Gusev, H.- C. Lu, E. L. Garfunkel, T. Gustafsson and M. L.Green, IBM J. Res. Develop. **43**, 265 (1999).
2. T. Hori, IEEE Trans. Electron Devices **37**, 2058 (1990).
3. S. Okada and H. Matsumura, Jpn. J. Appl. Phys. **30**, 3774 (1997).
4. H. Matsumura and H. Tachibana, Appl. Phys. Lett. **47**,833 (1985).
5. H. Matsumura, Jpn. J. Appl. Phys. **37**, 3175 (1998).
6. A. Izumi and H. Matsumura, Appl. Phys. Lett. **71**, 1371 (1997).
7. A. Izumi, A. Masuda, S. Okada and H. Matsumura, *Inst. Phys. Conf. Ser. No 155: Chapter 3* , p.343 (1997).
8. A. Izumi, A. Masuda and H. Matsumura, Thin Solid Films **343/344**, 528 (1999).
9. A. Izumi, H. Sato, S. Hashioka, M. Kudo and H. Matsumura, Microelectronic Engineering (1999) *to be pubulished*.
10. R. I. Hedge, P. J. Tobin, K. G. Reid, B. Maiti and S. A. Ajria, Appl. Phys. Lett. **66**, 2882 (1995).
11. E. C. Carr and R. A. Buhrman, Appl. Phys. Lett. **63**, 54 (1993).
12. L. M. Terman, Solid State Electron **5**, 285 (1962).

Integration Issues in the FEOL

Mat. Res. Soc. Symp. Vol. 611 © 2000 Materials Research Society

Integration Challenges For Advanced Salicide Processes And Their Impact On CMOS Device Performance

K. WIECZOREK[1], M. HORSTMANN[1], H.-J. ENGELMANN[1], K. DITTMAR[1], W. BLUM[1], A. SULTAN[2], P. BESSER[2], A. FRENKEL[2]

[1] AMD Saxony Manufacturing GmbH, Postfach 110110, D-01330 Dresden, Germany
[2] AMD-Motorola Logic Alliance, 3501 Ed Bluestein Blvd., Austin, TX 78721

ABSTRACT

$CoSi_2$ has emerged as the silicide of choice for $0.18\mu m$ CMOS technologies and below. Robustness and scaling-performance of an integrated $CoSi_2$-module, however, is shown to critically depend upon careful optimization of each individual process-step. The impact of surface-preparation, capping layer, initial Co-thickness and thermal processing will be discussed. The scalability of an optimized process meeting all major requirements for application to ULSI devices is demonstrated for gate-length down to 60nm.

INTRODUCTION

With minimum gate features for advanced CMOS-technologies approaching the sub-100nm range, reliable low-resistive silicide formation is a continuos manufacturability-challenge. For it's compatibly with major integration requirements [1], $TiSi_2$ has been the most widely applied material to ensure both low gate-delays and low-ohmic contacts to source and drain of CMOS devices. Decreasing gate geometries, however, stressed the downsides of the nucleation-and-growth dominated C49 to C54 conversion required for low-resistive $TiSi_2$-formation. Though high-dose preamorphization ion-implantation (PAI) or metal doping of the initial Ti-layer have been shown to significantly extend the scaling-performance of $TiSi_2$ [2], $CoSi_2$ has emerged as a

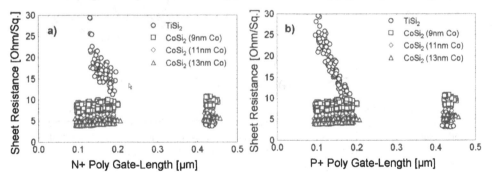

Fig. 1: Silicided gate sheet-resistance vs. gate-length comparing $CoSi_2$ formed from 9nm to 13nm initial Co-thickness against a high-energy preamorphization (PAI) $TiSi_2$-process for a) n^+-doped polysilicon and b) p^+-doped polysilicon.

replacement with the introduction of 0.18µm process technologies into volume manufacturing. While CoSi$_2$ features an excellent narrow line sheet-resistance scalability down to well below 100nm on both n$^+$- and p$^+$-doped polysilicon even for aggressively scaled Co-layers (Fig. 1), it's compliance with ultra-shallow junctions depends strongly on the details of the formation sequence [3]. Combining low gate- and source/drain-sheet-resistance requirements with the necessity to maintain ultra-shallow junction integrity requires subtle balancing of all processes within an integrated CoSi$_2$-module.

SURFACE PREPARATION

Since Ti has the capability of even reducing thin residual SiO$_2$ surface layers during the 1st RTA process due to it's high chemical reactivity, surface-cleaning has traditionally not been a major concern for reliable TiSi$_2$ formation. Even low-level surface contamination, however, has a profound effect on the control of the electrical and morphological characteristics of integrated CoSi$_2$-layers. Fig. 2 shows the impact of surface cleaning conditions on diode-leakage distributions. All samples were cleaned using a conventional RCA-chemistry and a subsequent

Fig. 2: Diode leakage as a function of pre-deposition surface treatment. Samples cleaned without any in-situ RF sputter etch process prior to Co-deposition suffer from intolerable diode-leakage distributions due to uncontrolled CoSi$_2$-morphology. A decreased diode leakage current is achieved with an additional in-situ RF sputter etch process.

dilute HF process for residual oxide removal and hydrogen saturation of the silicon surface so as to avoid immediate reoxidation. The samples were then processed within one hour through the Co deposition process. While one subset underwent Co deposition without any additional surface treatment, the two remaining subsets were in-situ sputter cleaned by Ar RF-sputtering to remove a target oxide thickness of 5nm and 10nm, respectively. While increased sputter-etch time reduces the high-leakage tail of the diode-current distributions, the non sputter-etched samples suffer from uncontrollable diode leakage with a range of four orders of magnitude. Cross-sectional transmission electron microscopy (TEM) reveals both a high $CoSi_2$ surface- and $CoSi_2$ - Si interface-roughness in conjunction with a very granular film-morphology. From large-area auger electron spectroscopy (AES) analysis as shown in Fig. 3, carbon has been identified as the primary residual contaminant after wet-chemical cleaning, particularly over active areas and polysilicon. The carbon surface concentration is significantly reduced over field-oxide. In-situ Ar sputter cleaning is effective in reducing silicon-carbonate or silicon-hydrocarbonate surface contaminations below the AES detection limit. The improved surface cleanliness promotes effective Co-Si interdiffusion and results in significantly reduced $CoSi_2$-induced junction leakage.

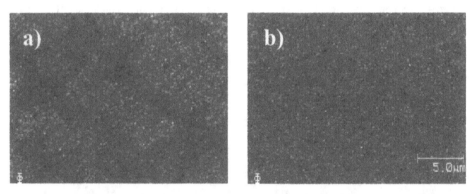

Fig. 3: AES mapping for detection of carbon surface contaminations. Carbon appears light on a dark background. From a) it is shown that carbon is mostly concentrated on active areas and on polysilicon lines after wet-chemical cleaning in a conventional RCA-chemistry. After RF sputter etching (b), no carbon is detected on the wafer surface.

The conditions of the RF sputter clean may, however, significantly impact $CoSi_2$-formation on active areas between dense polysilicon lines (Fig. 4). High-bias sputter clean conditions are required to yield a coherent $CoSi_2$ layer, whereas low-bias results in a discontinous film. Complete $CoSi_2$-formation, however, is essential to reliably controlling parasitic device-resistances in high-density circuit areas. Decreasing polysilicon pitches associated with design rule shrinks will increase the challenges associated with well-controlled silicide formation in high aspect-ratio environments. As shown in Fig. 5, redeposition of sidewall-spacer dielectric material to a thickness of 1.6nm suffices to partially block Co-diffusion into the underlying substrate. High-bias process conditions support the removal of the dielectric, but create a thin amorphous surface layer. Optimized process conditions will have to balance effective surface cleaning and control of sputter-etch induced radiation damage.

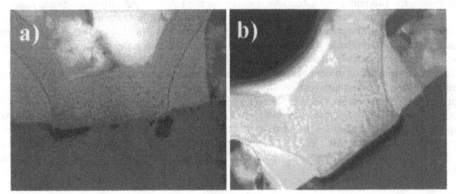

Fig. 4: CoSi$_2$ on active areas between dense gates for a) low-bias and b) high-bias RF sputter etch conditions.

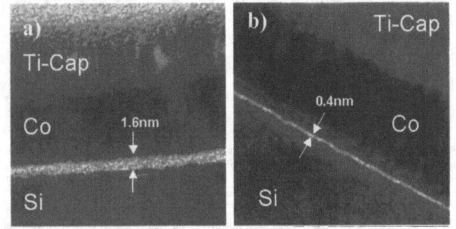

Fig. 5: Cross-sectional TEM of active areas between dense gates after Co and Ti-cap deposition for a) low-bias and b) high-bias in-situ RF sputter etch conditions.

IMPACT OF CAPPING LAYER

As given in Fig. 6, the choice of the capping-layer fundamentally modulates the sheet-resistances of integrated CoSi$_2$- layers. Even under well-controlled ambient-conditions within the RTA-system and thoroughly optimized purge-cycles, silicidation of uncapped Co-layers will result in vastly degraded electrical performance metrics. Desorption of oxygen-containing gaseous impurities from the wafer during temperature ramp-up and subsequent desorption has been identified as an uncontrolled oxygen-source with potential to oxidize as-deposited Co-layers during the 1st RTA cycle [4]. Therefore an in-situ deposited capping layer on top of the initial CoSi$_2$-film is indispensable for reliable and manufacturable CoSi$_2$-formation. While TiN-capping results in a relatively low wide-line sheet-resistance, as show in Fig.6, the use of a Ti-cap will largely increase narrow-line scaling-potential of CoSi$_2$-layer for application to devices with gate-length below 100nm. The superior performance Co-Ti process over the Co-TiN process has been associated with the capability of Ti-films to getter desorbed gaseous contaminations during the 1st RTA process [5,6,7] and to the ability of Ti capping material to reduce interfacial SiO$_2$ layers between Co and Si.

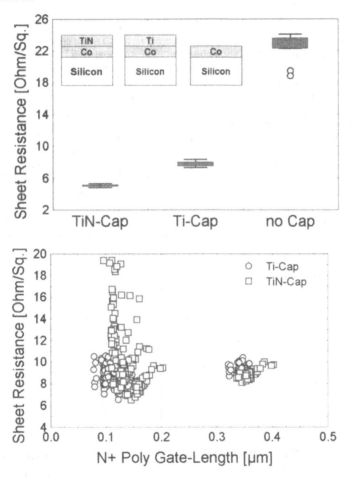

Fig. 6: Sheet-Resistance of CoSi$_2$ formed from constant Co-thickness as a function of capping layer and linewidth-dependence of CoSi$_2$ formed from Ti- and TiN-capped Co.

AES-depth profiles of Ti-capped Co-layers after 1st RTA (Fig.7a) confirm, that any oxygen contamination is effectively gettered within the Ti-film and thus separated from underlying CoSi. Futhermore, a distinct interdiffusion of Co and Ti is observed [8]. This will result in an effectively reduced amount of Co available for transformation into CoSi and thus an increased final CoSi$_2$ sheet-resistance as compared to a TiN-capped process. With the Ti-cap effectively inhibiting the oxidation of the Co, only small traces of oxygen are identified at the surface of the CoSi$_2$-layer after the 2nd RTA process. Note that the oxygen-level at the CoSi$_2$-Si interface is within the background noise-level (Fig.7b).

From cross-sectional TEM, as shown in Fig.8, a smoother CoSi$_2$-Si interface is observed for films formed from TiN-capped Co (Fig.8a) as compared to Ti-capped Co. Both samples were subject to the same surface-preparation and thermal treatment at 1st RTA and 2nd RTA; however the initial Co-thickness of the Ti-capped sample was adjusted to yield a comparable CoSi$_2$-thickness and sheet-resistance as the TiN-capped sample. For investigation of the underlying mechanism, high resolution elemental maps were taken from both samples by energy-filtering TEM (EFTEM). Fig.9a-d show a brightfield image and the corresponding Si-, Ti- and Co-elemental maps. While Si and Co are evenly distributed throughout the entire film, no Ti-signal is detected. The layer futhermore features a very uniform orientation.

In contrast to this observation, the CoSi$_2$-layer formed from Ti-capped Co displays a high Ti-concentration along the grain-boundaries as well as Ti-traces at both CoSi$_2$-surface and the CoSi$_2$-Si-interface (Fig.9e-f). With Ti occupying the main Si-diffusion path required for CoSi-to-CoSi$_2$ transformation during the 2nd RTA process, interface roughness modulation as a response to the choice of the capping layer is consistent with a grain boundary diffusion dominated silicide conversion process. The presence of Ti in CoSi$_2$-layers formed from Ti-capped Co has also been attributed to improved agglomeration-resistance and thus a higher thermal stability of CoSi$_2$ thin films [9].

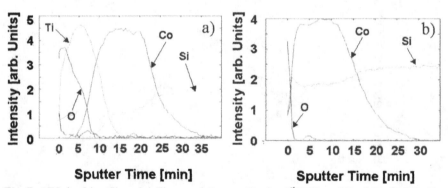

Fig. 7: AES depth profiles of a) Ti-capped Co over Si after 1st RTA and b) of the final CoSi$_2$ layer after 2nd RTA.

THERMAL PROCESSING

Continuous lateral scaling of CMOS gate-length below 100nm are associated with an increasing demand for both laterally and vertically abrupt ultra-shallow junctions with high dopant-concentrations in order to control short channel effects. Achieving and maintaining junction-integrity throughout a process-flow therefore is of increasing importance in order to control parasitic off-state currents. These two-dimensional scaling considerations represent a continuous challenge for balancing sheet-resistance and junction-integrity requirements.

The temperature of the 2nd RTA process has a significant impact on both CoSi$_2$-Si interface roughness, as show in Fig.10.

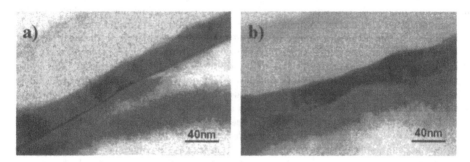

Fig. 8: Cross-sectional TEM of CoSi$_2$ layers formed from a) TiN-cappped Co and b) Ti-capped Co.

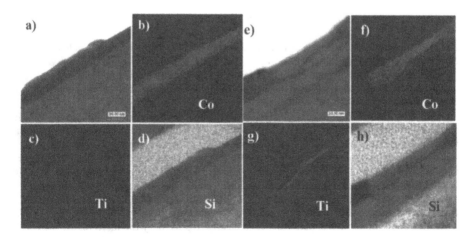

Fig. 9: Cross-sectional brightfield TEM and corresponding EFTEM elemental maps of CoSi$_2$-layers formed from TiN-capped Co (a-d) and Ti-capped Co (e-h).

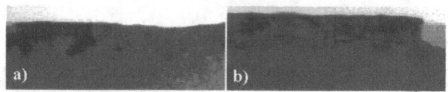

Fig. 10: Cross-sectional TEM of $CoSi_2$-layers formed from Ti-capped Co at a 2nd RTA temperature of a) T_2 and b) $T_2+100°C$.

Within a range of 100°C, the interface-roughness of $CoSi_2$ formed from Ti-capped Co can be significantly improved. The sensitivity of both junction-leakage and sheet resistance on the temperature of the 1st and 2nd RTA process as well as on the initial Co-thickness can be evaluated from Fig.11. Within a thickness range of 9nm to 13nm for the initial Co-layer and 1st and 2nd RTA temperature-ranges of 20°C and 50°C, respectively, Co-thickness is identified as the first order effect impacting sheet-resistance, followed by the 1st RTA temperature. This is in accordance with the assumption, that the thickness of the final $CoSi_2$-layer is mostly impacted by these two factors. For the 2nd RTA temperature range explored in this experiment, no significant impact on the final sheet-resistance is seen. In contrast to the sheet-resistance observations, the 2nd RTA temperature does significantly impact the diode leakage distributions. It's impact does, however, clearly depend on the initial Co-thickness and the 1st RTA, as shown in Fig.11. For both high and low 1st RTA temperature, the junction leakage distributions yield a significant improvement with increasing 2nd RTA temperature in case of 13nm initial Co-thickness. The same effect is observed for an initial Co-thickness of 9nm in combination with high-temperature 1st RTA, while a reversed effect occurs when 9nm Co and low temperature 1st RTA are combined, yielding degraded junction leakage for increased 2nd RTA temperature. This highlights the necessity to individually optimize 1st and 2nd RTA temperature for a given initial Co-thickness.

SCALING POTENTIAL AND CMOS DEVICE PERFORMANCE

The robustness and scaling performance of a Ti-capped $CoSi_2$ module carefully balanced to yield both controlled sheet-resistance and low junction leakage is shown in Fig.12. Each gate-length has been individually evaluated by fully automated CD-measurements after gate-patterning and resist-removal. Constant sheet-resistances down to 60nm have been achived, thus confirming the applicability to sub-100nm technologies.

Superior CMOS-device performance has been achieved compared to a high-dose PAI $TiSi_2$ reference-process in a 180nm 0.18V CMOS technology. The improved drive-current strength of the $CoSi_2$-process is related to a significant reduction of the S/D-resistance associated with the lack of a post junction-anneal high-dose PAI process. The lower $CoSi_2$ sheet resistance at minimum gate-geometries in conjunction with decreased S/D-resistance provide a scaling potential down to at least 60nm. Even in 0.18μm technologies, manufacturability benefits from reduced $V_{T,sat}$ roll-off and a 30nm drain-induced barrier lowering (DIBL) improvement of PMOS devices processed with an optimized $CoSi_2$ process. Besides improvements in AC-characteristics, ring-oscillator performance is clearly improved by the more robust narrow-line sheet resistances.

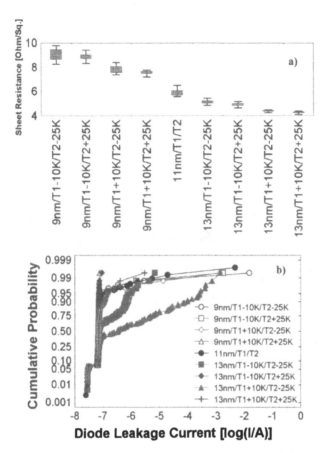

Fig. 11: Sheet resistance (a) and diode leakage (b) of CoSi$_2$-layers formed from Ti-capped Co as a function of initial Co-thickness, 1st RTA temperature T$_1$ and 2nd RTA-temperature T$_2$.

CONCLUSIONS

An integrated CoSi$_2$-module has been evaluated with respect to pre-deposition surface treatment, initial Co-thickness, capping-layer and thermal processing. It has been shown, that careful targeting of each individual process is required to simultaneously yield low sheet-resistances down to 60nm and well-controlled ultra-shallow junction reverse-bias leakage.

Ti-capping has shown to significantly improve the scaling-performance of the process over TiN-capping. Increased interface-roughness associated with the integration of a Ti-capping layer has been correlated to Ti-segregation along the grain-boundaries of CoSi$_2$ by means of cross-sectional EFTEM analysis.

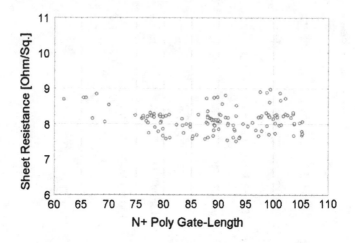

Fig.12: Sheet resistance as a function of gate-length for $CoSi_2$ formed from Ti-capped Co. The length of each gate has been examined individually by means of automated CD-SEM evaluation.

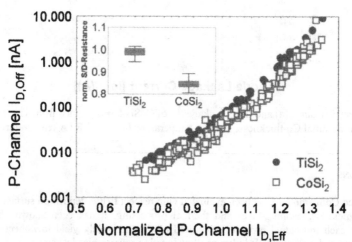

Fig. 13: Improved p-channel transistor drive current at a given off-state leakage has been achieved with a Ti-capped $CoSi_2$-process as compared to a reference high-dose PAI $TiSi_2$-process. The optimized device characteristics are associated with decreased source-drain resistances.

REFERENCES

[1] S.P. Murarka: Silicides for VLSI-applications, Academic Press Inc., Orlando (FL), 1983

[2] J.A. Kittl, Q.Z. Hong, H. Yang,N. Yu, M. Rodder, P.P. Apte, W.T. Shiau, C.P. Chao, T. Breedijek, M.F. Pas; Mat. Res. Soc. Symp. Proc., **525**, p.331-336, 1998

[3] K. Goto, A. Fushida, J. Watanabe, T. Sukegawa, Y. Tada, T.Nakamura, T. Yamazaki, T. Yamazaki, T.Sugii; IEEE Trans. El. Dev., **46** (1), p.117-138, 1999

[4] E. Kondol, K. Maex; Mat. Res. Soc. Proc., **470**, p.245-251, 1997

[5] K. Maex, A. Lauwers, P. Besser, E. Kondoh, M. de Potter, A.Steegen; IEEE Trans. El. Dev., **46**(7), p.1545-1550, 1999

[6] K. Maex, E. Kondoh, A. Lauwers, A. Steegen, M. de Potter, P. Besser, J. Proost; Mat. Res. Soc. Symp. Proc., **525**, p. 297-306, 1998

[7] P. Besser, A. Lauwers, N. Roelandts, K. Maex, W. Blum, R. Alvis, M. Stucchi, M. de Potter; Mat. Res. Soc. Symp. Proc., **514**, p. 375-380, 1998

[8] K. Maex, Materials Science and Engineering, **R11**(2-3), 1993

[9] D.K. Sohn, J.-S. Park, B.H. Lee, J.-U. Bae, K.S. Oh, S.K. Lee, J.S. Byun, J.J. Kim; IEDM Tech. Dig., p.1005-1008, 1998

ACKNOWLEDGEMENT:

The authors would like to thank V. Kahlert and D. Rowland for their outstanding support. They are futhermore indebted to R. Stephan and F.N. Hause for valuable discussion.

Mat. Res. Soc. Symp. Vol. 611 © 2000 Materials Research Society

Development of a Post-Spacer Etch Clean to Improve Silicide Formation

Edward K. Yeh[*], Samit S. Sengupta, and Calvin T. Gabriel[**]
Philips Semiconductors, San Jose, CA.

[*] Presently at Intel Corporation, Santa Clara, CA
[**] Presently at AMD, Sunnyvale, CA.

ABSTRACT

The formation of a uniform, low-resistance silicide is dependent on the contamination level of the silicon surface upon which the Ti or Co is deposited. Contamination may be in the form of embedded carbon and fluorine residues from previous processing steps, such as the LDD spacer etch. The extent to which the spacer etch contributes impurities can be determined by making contact angle measurements; that is, measurement of the angle formed by a droplet of water placed on the wafer surface. A clean silicon surface provides a contact angle of $> 70°$, whereas a contaminated silicon surface produces contact angles down to only a few degrees. By using contact angle as a metric, a post-spacer etch clean was developed to remove impurities from the silicon surface without significant silicon loss. The clean procedure entails the use of a CF_4/H_2O plasma to treat the wafer directly after spacer etch. The clean process was then evaluated by the formation of a blanket Ti silicide followed by sheet resistance measurement. By using this cleaning process, a Ti silicide sheet resistance comparable to that formed on virgin silicon wafers was obtained on wafers that experienced the spacer etch. By comparison, wafers that were subjected to the spacer etch but not the post-spacer clean yielded a higher sheet resistance. Additional study of the post-spacer etch clean process revealed that oxide on the wafer (even the small amount formed during an O_2-based resist ash process) can prevent the CF_4/H_2O plasma from cleaning the surface, perhaps by blocking H in the cleaning plasma from extracting C and other contaminants from the silicon surface. Co silicide formation is even more sensitive to surface impurities than Ti silicide. Using the same post-spacer etch clean, uniform, low-resistance Co silicide formation has been demonstrated.

INTRODUCTION

The reaction between Co or Ti and silicon to form a low-resistance silicide is very sensitive to the condition of the silicon surface involved. Contamination from previous processing steps, such as spacer etch, inhibits the uniform formation of a low-resistance silicide. For example, contamination from the spacer etch process may be in the form of carbon, fluorine, and/or fluorocarbon residues embedded within the silicon surface. In the spacer etch process, a conformal dielectric film is anisotropically etched in a plasma to leave spacers next to the gate electrodes. The etch process typically uses fluorocarbon gas species such as CF_4 and CHF_3 to enact the etching process. When the silicon of the gate electrode and the diffusion areas is exposed to the etch chemistry, it becomes embedded with carbon and fluorine species from the plasma. X-ray photoelectron spectroscopy (XPS) analysis of a silicon surface after exposure to the spacer etch plasma shows the presence of these species (Figure 1). During subsequent salicide steps, these contaminants hinder formation of a uniform, low-resistance silicide. A typical solution is to again use a fluorocarbon chemistry to attempt to etch away the

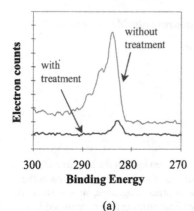 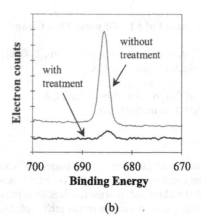

(a)

(b)

Figure 1. (a) C (1s) (peaks indicate C-C, C-O, and C-F_x species) and (b) F (1s) (peak indicates F-F species) XPS data from analysis of a silicon surface exposed to the oxide spacer etch chemistry (without treatment) and from analysis after the CF_4/H_2O clean process [2].

contaminated silicon layer. However, this results in additional silicon loss from the top of the gate electrode as well as from the exposed silicon areas. In light of the trend towards shallow junctions, devices will not be able to tolerate much of this silicon loss. Also, the use of another fluorocarbon plasma to remove the contaminated layer invariably leaves its own residues.

Contact angle measurement using deionized water (DI) droplets is a simple technique used to gauge the contamination level of a Si surface. The contact angle is defined as the angle between the horizontal plane at the base of the droplet and the drop profile near the edge of the drop through the droplet. These contacts angles exhibit the same sensitivity to silicon surface condition as the Ti silicide sheet resistance. When the surface has become embedded with residues, a droplet of DI water forms only small (< 10°) contact angles. However, when the silicon is clean, the contact angles obtained are much larger (> 60°). The goal is thus to obtain a high contact angle (i.e., a clean silicon surface) after spacer etch without much silicon loss.

Table I gives the results for earlier work done at 125°C in a Lam 9600SE DSQ chamber [1]. The data shows that by using CF_4 alone, the resulting high etch rates remove the contaminated layer, but also a great amount of the gate, isolation, junction, and spacer. This clean surface then provides a good (high) contact angle. The use of either H_2O or O_2 alone will not cause much etching, but does not remove any contamination as evidenced by the poor (low) contact angles. However, the combination of both CF_4 and H_2O gives the best of both worlds: low oxide and poly etch rates while providing for a good contact angle.

The use of contact angle (and associated blanket Ti silicide sheet resistance) measurements allowed a more efficient development of a post-spacer etch clean procedure to remove contaminants without significant silicon loss.

Table I. Early data on the development of a post-spacer etch clean [1].

Measurements	180 sccm CF_4	300 sccm H_2O	200 sccm O_2	180 sccm CF_4/ 300 sccm H_2O
Contact angle	58°	2°	4°	54°
Oxide etch rate (Å/sec)	11.36	0.04	0.04	0.08
Poly etch rate (Å/sec)	12.90	0.03	0.02	0.23

EXPERIMENTAL

200-mm diameter p-type Si wafers were used for the present experimental study. For conducting contact angle measurements, a blanket LPCVD TEOS oxide was first deposited on the wafers. The oxide layer was then anisotropically etched down to the silicon surface using a conventional CF_4/CHF_3-based chemistry in an RF diode plasma etch system to simulate a typical oxide spacer etch and subsequent overetch. The wafers were then subjected to various CF_4/H_2O plasma conditions in the ICP chamber of a Mattson Aspen Strip System. The water vapor (H_2O) feed was delivered at 70°C. A 30 min chilled DI water/ozone mixture followed by a 60 sec 100:1 HF dip completed processing. Within 1 hr of being wet cleaned, contact angles of droplets of DI water on these wafers were measured. Contact angles were measured using a MicroVu 500 HP Optical Comparator. This tool is not a true goniometer (an instrument for measuring contact angles), but was adequate for the purpose of experimental measurements, and will henceforth be referred to as the 'goniometer.' Approximately 4 to 5 μl water droplets were dispensed using a calibrated pipette.

Variations in the measurement of the contact angle can be caused by initial spreading of the droplet (reducing the measured contact angle), and additional spreading caused by motion of the stage during focus, human error, an inclined goniometer stage, and/or an inclined table (increasing the contact angle for the lower side of the droplet). To compensate for the latter sources of error, the contact angle values discussed herein are an average of the angles measured on the left and right side of the droplet.

Blanket Ti silicide wafers used in the evaluation of the clean process were separately obtained using identical deposition and etch processes as described above. After the CF_4/H_2O plasma treatment, the wafers were then subjected to the spacer wet strip process. This strip process entailed the use of a sulfuric-ozone clean (H_2SO_4 at 90°C with ozone bubbled through it), a 100:1 HF dip, an SC1 clean (5:1:0.5 H_2O:H_2O_2:NH_4OH at 50°C), and an SC2 clean (5:1:1 H_2O:H_2O_2:HCl at 50°C). The next step was the Ti deposition HF-based pre-clean, which was immediately followed by a 350 Å Ti deposition, a 720°C rapid thermal anneal (RTA) to form C49-phase Ti silicide (RTA-1), an SC1 clean (5:1:1 H_2O:H_2O_2:NH_4OH at 40°C) to remove TiN and unreacted Ti, and a final 850°C RTA (RTA-2) to transform the C49-phase to the low-resistance C54-phase. The blanket sheet resistance of the wafer was then measured using a CDE four-point probe.

RESULTS AND DISCUSSION

A. TI SILICIDE RESULTS

Using the 0.6:1 CF_4:H_2O gas ratio in Table I and keeping with standard recipe elements for the Mattson Aspen Strip System, the initial process conditions established were 900 W RF, 1.2 Torr, 360 sccm CF_4, and 600 sccm H_2O with a 250°C platen temperature. The goal was to obtain the same contact angle on a wafer that had experienced the spacer etch as a virgin silicon wafer. Typical contact angles measured on virgin silicon wafers after a 60 sec 100:1 HF dip were about 70 to 75 degrees. Following the procedure outlined previously for contact angle measurements, the CF_4/H_2O RF time was varied to determine the appropriate process times (Figure 2). Figure 2 shows that as the CF_4/H_2O RF time increases, the contact angle increases as well up to about 60 sec, beyond which the contact angle values plateau within the range

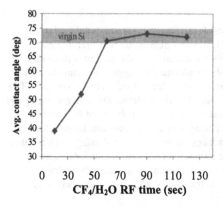

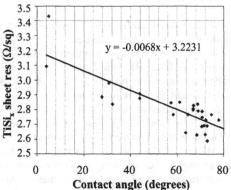

Figure 2. Average contact angle as a function of CF_4/H_2O RF time. The gray band indicates the range of contact angles typically measured on virgin silicon wafers.

Figure 3. Correlation of $TiSi_x$ sheet resistance with contact angle.

measured for virgin silicon. This indicates that after 60 sec, the previously etched surface was restored back to an uncontaminated state. To allow sufficient process margin, a time of 90 sec was chosen for further experiments. To check for silicon and oxide loss, amorphous silicon and oxide etch rates were measured for the above process. Both etch rates were approximately 0.18 Å/sec, indicating a loss of only 16 Å for the 90-sec process.

Next, the measured contact angles were correlated with blanket Ti silicide sheet resistance. Figure 3 reveals a roughly linear relationship and validates the use of contact angles as a measurement of the surface contamination.

Using the 90-sec process described earlier on wafers that experienced the spacer etch, a contact angle of 73° and a sheet resistance 2.59 Ω/sq (1.59% 1σ) were obtained. These numbers are comparable to that obtained on virgin silicon wafers, e.g., a 71° contact angle and 2.69 Ω/sq (1.75% 1σ) sheet resistance. XPS analysis of a treated wafer revealed that the carbon and fluorine signals present after etch were greatly reduced by the CF_4/H_2O clean (Figure 1). For comparison, wafers that were subjected to the oxide spacer etch, but not to the CF_4/H_2O treatment, gave a 28% higher sheet resistance of 3.33 Ω/sq.

Additional study of this post-spacer etch clean process revealed that oxide on the wafer, even the small amount formed during an O_2-based resist ash process, can prevent the CF_4/H_2O plasma from removing contaminants from the silicon surfaces. For example, a 15 sec O_2 ash prior to the 90 sec CF_4/H_2O plasma clean resulted in a poor 4° contact angle. Carrying this through the Ti silicide process resulted in a poor sheet resistance of 3.09 Ω/sq. Increasing the O_2 ash time to 45 sec (to grow more oxide) also gave a low contact angle (5°), but a higher sheet resistance (3.43 Ω/sq). To explore whether oxide growth was limiting the effectiveness of the clean process, 2% CF_4 was added to the 15 sec O_2 ash to provide an oxide-etching component to remove any oxide that would otherwise form. The result of this was a dramatic increase in contact angle to 69° and a reduction in sheet resistance to 2.79 Ω/sq. A follow-up experiment involving a 45 sec O_2 ash, followed by a 60 sec 100:1 HF dip to remove oxide, and then the standard 90 sec CF_4/H_2O clean resulted in a sheet resistance of 2.61 Ω/sq (comparable to that obtained on virgin silicon). These experiments verify the ability of oxide to block the post-spacer etch clean process. As expected, placing the O_2 ash after the CF_4/H_2O plasma did not seriously degrade Ti silicide sheet resistance, e.g., a 15 sec O_2 ash and a 45 sec O_2 ash after 90 sec of the CF_4/H_2O plasma resulted in resistances of 2.63 Ω/sq and 2.85 Ω/sq, respectively. The higher sheet resistance associated with the 45 sec post-CF_4/H_2O ash may be due to increased oxide growth that is not fully removed prior to Ti deposition and acts as a barrier to silicidation.

B. THE MECHANISM OF THE CF_4/H_2O PLASMA

The use of fluorine in the plasma would seem to indicate etching of the silicon surface to remove surface contamination. However, from previous oxide and silicon etch rate measurements, it is seen that the CF_4/H_2O plasma does not provide for much etching of those films. By focusing an optical emission spectrometer (OES) on CF_4, H_2O, and CF_4/H_2O plasmas,

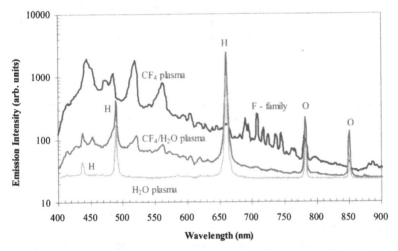

Figure 4. Comparison of optical emission spectra for CF_4, H_2O, and CF_4/H_2O plasmas [2].

an explanation is available (Figure 4). In the CF_4 plasma, the OES spectrum shows several peaks between 680 and 780 nm corresponding to a family of fluorine signals. On the other hand, the H_2O plasma displays mainly hydrogen and oxygen peaks. The combination of the two in the CF_4/H_2O plasma shows that the family of fluorine peaks is greatly suppressed, probably due to scavenging by free hydrogen. As the concentration of free fluorine is diminished, the etch rate resulting from the CF_4/H_2O plasma diminishes as well.

The reaction between Co and silicon to form Co silicide is even more sensitive to the presence of surface contamination than Ti silicide. However, the use of the same 90 sec CF_4/H_2O plasma clean still allowed for the formation of approximately 600 Å of a uniform, low-resistance Co silicide atop the gatestack and over the active areas. Quantitatively, with a 120 Å Co/80 Å Ti blanket film stack and no spacer etch exposure, a sheet resistance of 4.54 Ω/sq (0.76% 1σ) with no pre-sputter etch was obtained. On a 0.20-μm n-diffusion area, the same process flow as the blanket wafers described above provided Co silicide with a sheet resistance of 4.74 Ω/sq (6.30% 1σ) with no pre-sputter etch. [Additional experiments with the use of a pre-sputter etch did not provide any further improvement in sheet resistance.] The resistances measured on the 0.20-µm diffusion areas (that experienced the spacer etch) compare well to those obtained on the blanket wafers. The relatively higher (as compared with the blanket silicide) nonuniformities may be attributable to variations in the small CD size of the measured diffusion area.

CONCLUSIONS

The use of a CF_4/H_2O plasma clean can remove carbon and fluorine contamination left within the silicon by the spacer etch process. The removal of these residues leaves a clean silicon surface conducive to the formation of a uniform, low resistance Ti or Co silicide. Without the use of this surface treatment, Ti silicide formation is hindered, yielding a 28% higher blanket sheet resistance. The mechanism by which the CF_4/H_2O plasma accomplishes the cleaning involves the scavenging of fluorine by atomic hydrogen, thus limiting the etch rates, yet still removing carbon and fluorine etch residues. Further study of the process revealed that oxide on the wafer surface can prevent the removal of those same residues, resulting in poor silicide formation. This post-spacer etch clean has also proved applicable to Co silicide.

ACKNOWLEDGMENTS

The authors would like to thank past and present members of the VLSI Technology Development group, especially Suzanne Monsees for her cross-section SEM work and Linda Leard and Charlotte Larin for their processing work. In addition, the authors would like to thank Vincent Melchior, Bob Guerra, Sai Mantripragada, Rene George, and Buffalo Zobel of Mattson Technology for their contributions.

REFERENCES

1. Unpublished results of X.-W. Lin, I. Harvey, H. Lee, and R. Solis while at VLSI Technology, Inc. See also U.S. Patent No. 5,895,245.
2. R. Solis, I. Harvey, and C. Gabriel, Proc. 6th Intl. Symp. on Semiconductor Manufacturing (1997).

Mat. Res. Soc. Symp. Vol. 611 © 2000 Materials Research Society

A TWO-STEP SPACER ETCH FOR HIGH-ASPECT-RATIO GATE STACK PROCESS

Chien Yu, Rich Wise*, Anthony Domenicucci
IBM Microelectronics, Semiconductor Research & Development Center, East Fishkill, NY 12533
*DRAM Development Alliance IBM/Infineon

ABSTRACT

A highly selective nitride etch was developed for gate stack spacer process in advanced memory programs. Based on methyl fluoride chemistry with better than 8:1 selectivity of nitride:oxide, this process exhibits minimal erosion to the underlying RTO thermal oxide for consistent diffusion ion-implant control. As the groundrule changed to 0.175um and below, a two-step etch scheme was employed to maintain the profile control in high-aspect-ratio structures. The stability and repeatability of the process is demonstrated in the SPC chart of the post etch FTA site measurement.

INTRODUCTION

As the semiconductor devices evolve rapidly into the deep sub-micron regime, the use of silicon nitride poly gate spacer has become ever more entrenched. Besides being a highly effective isolating material (compared to oxide), the nitride spacer also enhances the gate fringing field effect due to its high dielectric constant for improved device performance [1], particularly for the densely packed array area. The combined use of thick nitride cap layer and nitride spacer also provides the best protection for gate silicide during borderless contact etch from shorting to the subsequent diffusion contact fill in DRAM technologies.

The nitride spacer is typically made by an anisotropic dry etch of a conformally deposited silicon nitride layer around the gate stack after the poly re-oxidation. It is essential to select etch process with high selectivity to oxide and with minimal lateral etch. With high selectivity to oxide the nitride etch can be stopped on the screen (sidewall) oxide with minimal oxide erosion. This would eliminate the possibility of silicon damage in the diffusion area, and provide a robust and reproducible screen oxide for diffusion ion implant. With minimal lateral etch a sidewall spacer conformal around the gate stack with near as-deposited thickness next to silicide can be obtained for best isolation.

The unrelenting trend of increasing array density and decreasing effective gate-length results in a corresponding increase of aspect ratio of the gate stack over time. Toward the end of 0.35 um DRAM technology generation, we have developed a highly selective nitride spacer etch based on methyl fluoride chemistry to replace an older HBr/Cl$_2$ based process. In the subsequent generations we found that the increase of the gate trough aspect ratio resulted in a gradual increase of foot. This can be a potential issue if the spacer process is followed by diffusion ion-implant. The extent of the foot became severe enough in 0.175um generation that a two-step etch scheme was further developed to resolve the profile control issue, while preserving the desirable high selectivity etch as overetch step to soft-land on the screen oxide.

EXPERIMENTAL

Process development work was done in a commercial MERIE etcher. 3865Å emission was chosen for endpoint control. Both scanning electron microscopy (SEM) and transmission electron microscopy (TEM) were used for construction analysis. The latter was employed when high clarity images were required. Post etch sidewall oxide measurements were made in a thin film measurement tool in ellipsometry mode.

DISCUSSION

- **HIGHLY SELECTIVE NITRIDE ETCH**

The highly selective nitride spacer etch differs substantially from other selective etches in that selectivity is sought for silicon nitride over silicon dioxide (rather than the inverse), and in that typically no photoresist is available locally at the spacer location to enhance polymerization and hence selectivity.

Selectivity of the silicon nitride patterned spacer relative to the underlying thermal oxide may be achieved in one of two ways. In our original process, selectivity was achieved by chemical means: a chlorine/bromine chemistry was employed to minimize erosion of the oxide. Unfortunately, this chemical selectivity was highly dependent on the ratio of ion to neutral flux. As the plasma density and operating pressure were decreased to minimize aspect ratio dependent effects on smaller groundrule features, this favorable neutral to ion flux ratio was difficult to maintain. Additionally, the HBr/Cl_2 chemistry would rapidly attack any exposed silicon.

The methyl fluoride process, in contrast, utilizes polymerization as the mechanism for selectivity of nitride:oxide, although the mechanism is expected to be much different than that for the inverse (oxide:nitride) case. Oxide:nitride selectivity is achieved through the use of highly polymerizing fluorocarbon (e.g. C_4F_8), which tends to deposit a Teflon-like coating on all surfaces[2]. Those surfaces exposed normally to the plasma are heated locally by high ion energy and flux (i.e. high bias power and plasma density). This ion bombardment serves to form a "mixing layer", wherein the nonvolatile fluorocarbons are mixed with underlying silicon dioxide, and the high impact energy yields reactions of the sort:

$$CxFy + SiO_2 + Ion => n(CO + CO_2) + SiFz + Ion,$$

the ion may or may not participate chemically in the reaction. All products over oxide are oxidized and form volatile compounds, whereas on silicon nitride oxidation does not occur. The polymer chemistry may be tuned to provide for a greater sticking coefficient on silicon nitride than that of silicon oxide, further enhancing the selectivity.

The role of polymer deposition for silicon nitride:silicon oxide selectivity with methyl fluoride chemistry appears to be quite different. The methyl fluoride chemistry uses low plasma power, both to provide a thick, readily sputtered, hydrogen rich polymer, and to prevent high energy bombardment of oxide surfaces and subsequent mixing (and therefore volatilization) of this polymer. Selectivity is achieved by tuning the sticking coefficient of the deposited polymer to selectively deposit on the silicon dioxide (or silicon). In addition, the sputter yield for silicon dioxide is generally lower than that for silicon nitride, such that a low ion energy (bias power) is capable of providing a high yield of silicon nitride relative to silicon dioxide. Our methyl fluoride process yielded nitride:oxide selectivity in the > 8:1 range, consistent with previous selectivity obtained with HBr/Cl_2 at larger groundrules.

The methyl fluoride process, by design, is highly sensitive to polymer chemistry in the chamber, and careful attention must be paid to the chamber condition during the spacer etch. Any increase in the CFx content of the chamber can result in excessive footing of the spacer, since the sputter yield is designed to be low. As will be discussed below, reduction of this foot was achieved by minimizing the total etch depth during which the highly selective step was used. In our example, only the overetch of the spacer process required high (> 8:1) nitride to oxide selectivity, and we applied the selective step for approximately the final 50% of the etch to adjust for non-uniformities in etch rate and spacer thickness.

- **IMPLEMENTATION OF THE NITRIDE SPACER ETCH**

The implementation of the new spacer etch was driven by the groundrule reduction. It was also timed at the tail end of our 0.35um technology development cycle so the process change was verified to be transparent to the structure and device performance.

As the new nitride spacer etch was implemented in the .25um groundrule technology and beyond, a small undesirable feature was observed to evolve when single-step etch process was used. In the base of the spacer there is a small tapering or footing extended outward from the polygate. The extent of the foot appears to increase in the successive generation of technology cycles as shown in the SEM photos in Figure 1 below in next page.

The extent of the foot from a straight profile appears to correlate to the aspect ratio of the gate stack trough as summarized in the table below:

Technology	Spacing	Height	Aspect ratio	Size of foot
0.35 um	270 nm	415 nm	1.54	0 nm
0.25 um	180 nm	320 nm	1.67	6 nm
0.20 um	180 nm	350 nm	1.94	10 nm
0.175 um	120 nm	300 nm	2.5	13 nm

This apparent correlation can be understood as due to the increased constraint from the increasing aspect ratio of the gate stack trough. With less ion sputtering and lower pumping speed toward the bottom of the trough as well as decreased neutral species mass transfer, there is an increasing buildup of polymer at the base of the spacer as the etch progresses. This is not a significant problem in .25 um and .20 um generations as the diffusion ion implant was done before the spacer formation. However, in 0.175 um groundrule the integration sequence is reversed and we have to reduce the footing at the bottom of the spacer.

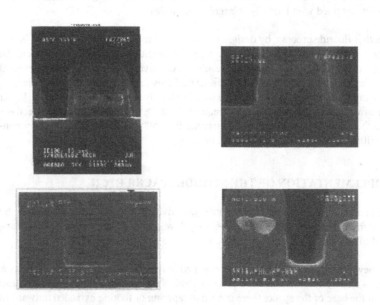

Figure 1: gate spacer of 0.35um groundrule (upper left), 0.25um (lower left), 0.20um (upper right), 0.175um (lower right) of single-step nitride spacer etch.

A two-step etch scheme was since designed to resolve the profile control issue. By starting the etch with a non-selective nitride etch with greater sputtering component (CHF$_3$/CF$_4$/Ar), finishing with the selective etch, the spacer foot was completely removed without any compromise of the favorable features of the new etch. The efficacy of this scheme is demonstrated in the TEM photos as shown in Figure 2. Using wafer from 0.175 um technology, with 30 nm as-deposited silicon nitride film, the single-step etch produced the spacer with a width 42 nm at the base of the poly gate due to the extended foot. The two-step etch produced a straight profile with 30 nm of spacer width matching the deposited thickness.

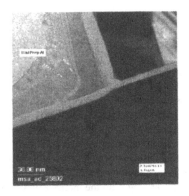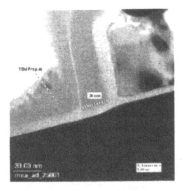

Figure 2: single-step etch with extended foot (left), two-step etch without foot (right)

- **SPACER ETCH PROCESS CONTROL**

To ensure process stability and repeatability in the product jobs, post spacer etch FTA site measurement has been generally used in the industry. The advantage of using highly selective nitride etch is demonstrated in the stable and unchanged screen oxide thickness before and after the etch. The correlation of the screen oxide and the FTA site oxide was verified with TEMs to be both at about 10nm. This close correlation enabled us to use the post etch measurement to monitor process integrity. A minute amount of polymer is left on the wafer after etch, the measurement is made after ozone clean. Figure 3 is the SPC chart of the post-etch FTA site measurement over six month period. The measurement is stabilized at 108A, in agreement with sidewall oxidation target. The two dash lines represent two etch chambers. There is no etch chamber difference for this measurement.

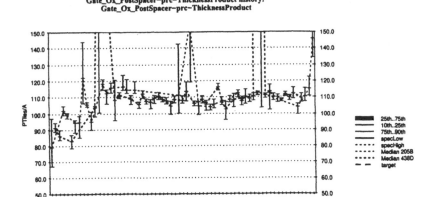

Figure 3: FTA site measurements after spacer etch over six month period for jobs from two etch chambers (thin and thick dash lines)

SUMMARY

We have reported the development of a new highly selective nitride etch and its implementation in the spacer etch for advanced DRAM technologies. Based on methyl fluoride chemistry, this new process is developed to replace a HBr/Cl$_2$ process of comparable nitride to oxide selectivity. The new process does, however, have far superior nitride to silicon selectivity. Unlike the other reported selective nitride etch using methyl fluoride chemistry [3], this process does not rely on high oxygen flow to provide oxide selectivity. Instead, hydrogen-rich polymer and lower sputtering are the keys to the high selectivity. With proper process design we are able to achieve virtually no RTO thermal oxide loss after etch . In 0.175 um groundrule and below, a two-step etch scheme was designed to resolve profile control issue arising from the increase of aspect ratio of the gate trough. The successful implementation of the two-step process was demonstrated in the TEM photos and the stable post etch FTA site measurements.

The authors would like to thank James Norum for managerial support, George Goth, Vic Nastasi and Mike Wise for helping qualify the new etch in DRAM programs, Delores Bennett for tool qualification support and Viraj Sardesai for integration support in the implementation of two-step etch process.

REFERENCE:

1. S. Wolf, in *Silicon Processing for the VLSI Era*, Vol. III (Lattice Press, 1995) pp. 634

2. T. Ichiki, Y. Chinzei, Y.Horiike, H. Shindo, N. Ikegami,m M. Sekine, 'Residence Time Effect on SiO$_2$/Si Selective Etching in High Density Fluorocarbon Plasma', 43th National Symposium, American Vacuum Society, Philadelphia, 1996

3. J.M. Regis, A.M. Joshi, T.Lill, M. Yu, "Reactive Ion Etch of Silicon Nitride Spacer with High Selectivity to Oxide", IEEE/SEMI Advanced Semiconductor Manufacturing Conference 1977, 252

Novel Silicide Processing

Mat. Res. Soc. Symp. Vol. 611 © 2000 Materials Research Society

IN SITU AND *EX SITU* MEASUREMENTS OF STRESS EVOLUTION IN THE COBALT-SILICON SYSTEM

G. Lucadamo*, C. Lavoie, C. Cabral Jr., R. A. Carruthers, and J.M.E. Harper

*Lehigh University, Dept. of Materials Science and Engineering, Bethlehem, PA 18015
IBM Research Division, Thomas J. Watson Research Center, P.O. Box 218, Yorktown Heights, NY 10598

ABSTRACT
   The biaxial stress in Co thin-films has been investigated *in situ* by measuring changes in substrate curvature that occurred during deposition and annealing. Films of Co, 35 to 500 nm in thickness, were deposited by UHV magnetron sputtering at room temperature on Si (100) and poly-Si substrates. Results show that during Co deposition the bending force increased linearly with film thickness; a signature of constant stress. In addition, the stress evolution during silicide formation was measured under constant heating rate conditions from room temperature up to 700 °C. The stress-temperature curve was correlated with Co_2Si, $CoSi$, and $CoSi_2$ phase formation using *in situ* synchrotron X-ray diffraction measurements. The room temperature stress for the $CoSi_2$ phase was found to be ~0.8 GPa (tensile) in the films deposited on Si (100) and ~1 GPa (tensile) on the films deposited on poly-Si. The higher tensile stress in the poly-Si sample could be a result of Si grain growth during annealing.

INTRODUCTION
   Cobalt silicide has been used to create low leakage junctions in sub-0.25 μm technologies [1]. Unlike Ti-silicide metallurgy, $CoSi_2$ forms easily in small structures as it does not exhibit polymorphism. In addition, $CoSi_2$ also has a slightly lower resistivity and forms at lower temperature than $C54$-$TiSi_2$. A complete optimization of the Co-salicide process should include an understanding of the stress changes that arise during the reaction. The formation of high stresses during processing could introduce defects, such as dislocations, that would be a hindrance to device operation [2].
   The experimental measurement of the intrinsic stress in vapor-deposited polycrystalline and epitaxial thin films has been the subject of many investigations over the past several years. Less common, but of significant technological importance, is the understanding of the stress changes that occur during solid state reactions in thin-film systems. This topic has been the subject of two reviews by d'Heurle [3-4]. For silicides formed under isothermal conditions the stress has been modeled by considering the sum of the instantaneous forces applied to a substrate by the growth of a compound phase minus relaxation during the reaction [5].
   Under constant heating rate conditions, extrinsic stress arises from differences in thermal expansion coefficients between the film and substrate, while intrinsic stress changes can develop from the formation of possibly several compounds. This type of measurement can be used to investigate the formation of phases over a range of temperatures, and thus provide important insight into the nature of the stress changes throughout a reaction sequence.
   During isochronal annealing, Co will react with Si initially forming Co_2Si at ~450 °C followed by $CoSi$ at ~ 500 °C, and finally $CoSi_2$ at ~600 °C. Several previous studies have investigated the stress change during silicide formation on Si (100) [2,6-7]. In this study, we present results describing the stress evolution of Co films during growth and subsequent annealing under constant heating rate conditions. The stress measurements were correlated with synchrotron X-ray diffraction (XRD) data to gain further insight into the nature of the stress changes present during silicide formation.

EXPERIMENT
Film Deposition
   Cobalt films with thicknesses ranging from 35 to 500 nm were deposited using an ultra-high vacuum (UHV) magnetron sputtering system with typical base pressures of $< 10^{-9}$ Torr. The

substrates used were single crystal Si (100) wafers, 125 mm in diameter with a nominal thickness of 625 μm. Some substrates were coated with a 150 nm poly-Si film on a thermally grown SiO_2 layer. Before loading into the vacuum system, the Si(100) and poly-Si substrates were dipped in a dilute HF solution to remove the native surface oxide. Depositions were carried out at room temperature using 3 mTorr of ultra-high purity Ar.

Stress Measurement

The biaxial stress present in the film during growth was measured using a k-Space Multi-beam Optical Sensor (KMOS) [8-9] and a KLA-Tencor Flexus FLX-2908 [10]. The KMOS unit is attached to the deposition system enabling *in situ* studies during film growth and annealing in a UHV environment. The FLX-2908 is a separate unit that measures wafer curvature using a scanning laser beam. It is integrated to a furnace and also gives *in situ* data during annealing.

In the present configuration, the KMOS unit uses a diode laser and beam splitter to obtain two parallel incident beams. The use of a beam splitter rather than the conventional etalon permitted increased spot separation on the wafer surface (~40 mm). The increased distance enhanced the sensitivity to changes in wafer curvature caused by film growth. When combined with a long substrate-to-detector distance (~1.7 m) and a megapixel digital camera, curvature changes corresponding to a stress-thickness resolution of ~ 2 MPa-μm were resolved for the substrates described above. This level of resolution is necessary to measure the stress changes typically associated with the growth of thin polycrystalline films used in the microelectronics industry. For example, typical growth stresses of a few hundred MPa can be detected in films as thin as 10 nm. Note that this level of resolution can be improved by about an order of magnitude by using 200 μm-thick wafers. In this study, system vibrations limited the use of thinner wafers. The stress-thickness data was acquired with and without substrate rotation. For measurements using rotation, an optical sensor mounted on the rotation assembly ensured that the spot spacing was measured once per rotation, at the same wafer position. This resulted in a sampling rate of one point every ~3.8 sec. For the static measurements, the sampling rate increased to 2 Hz. Using the relative change in spot spacing to determine the radius of curvature of the wafer, the biaxial modulus of the substrate (Si (100) = 180.5 GPa), and Stoney's equation, the stress-thickness product (bending force per unit width) is obtained as a function of film thickness [8].

As deposited stresses were obtained from the KMOS and KLA-Tencor Flexus FLX-2908 instruments. Changes in the stress-thickness during annealing were monitored by the KMOS and the FLX-2908 in UHV and forming gas, respectively. These anneals were conducted at 10 °C/min from room temperature to 700 °C. A second heating cycle was performed in the FLX-2908 without removing the wafer to ensure that the reaction had reached completion. In this instrument, thermal calibration experiments performed using a wafer with an attached thermocouple indicated that the maximum deviation between the wafer and set point temperature during the anneal was less than 8 °C.

In situ XRD

In situ synchrotron XRD studies were conducted using the IBM/MIT beamline X20C at the National Synchrotron Light Source located at Brookhaven National Laboratory. A multilayer monochromator provided an incident photon flux of about 10^{13} photons/sec. The equipment consisted of a position-sensitive detector formed of a linear array of diodes capable of measuring diffracted X-ray intensities from a 10° 2θ angular window [11]. Samples were heated in a purified N_2 or He ambient at heating rates of 10 °C/min or 180 °C/min from 100 to 1000 °C with a 1% accuracy in ramp rate and 3° C accuracy in temperature. Simultaneous four-point-probe resistance and elastic light scattering from both 5 and 0.5 μm lateral length scale were also acquired, but will not be discussed here. Further details about this technique can be found in previous publications [12-13].

RESULTS

Stress During Deposition

The curvature of the substrates was pre-measured prior to deposition using an average of at least 5 scans. The substrates were then HF dipped and loaded into the UHV system. Figure 1 shows a plot of bending force as a function of time for the growth of 100 nm of Co on Si (100).

A monotonic, linear increase of the bending force was observed producing a fairly uniform growth stress of ~ 600 MPa. The overall behavior was consistent with the deposition conditions given the low adatom mobility of Co [14-15]. The expected, initial compressive growth stress was not resolved in these experiments [14-15]. The evolution of stress during growth was also determined from several other 100 nm thick Co films deposited on Si (100) as well as a 35 nm thick film deposited on polycrystalline Si. The average growth stresses were ~720 MPa for the 100 nm film on Si (100). The increase in tensile stress ($\Delta\sigma$) after the shutter closes is most likely the result of substrate cooling.

The as deposited film stress was also measured using the FLX-2908. Following deposition, the average curvature of the wafer was determined using an average of at least 5 scans. The 100 nm-thick Co films resulted in an average tensile stress of 880 MPa. The 35 nm thick film had a stress of 780 MPa, and the 35 nm film deposited on poly-Si had a stress of ~ 1 GPa. The wafer-to-wafer variation in the stress measurement was greater for the poly-Si substrates. The results of the pre-anneal stress measurements are listed in table I.

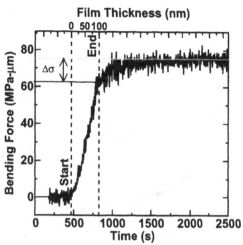

Figure 1: Bending force as a function of time for the deposition of 100 nm of Co on Si (100) at room temperature.

Table I: Stress ranges obtained for as-deposited Co films deposited on Si (100) and poly-Si.

Film	Substrate	Subst. Thick. (μm)	KMOS (MPa)	FLX-2908 (MPa)
100 nm Co	Si (100)	620	720	880
35 nm Co	Si (100)	620	—	780
35 nm Co	poly-Si/SiO$_2$	620	—	880, 1080

Stress During Annealing

Wafers were annealed at 10 °C/min to 700 °C. Measurements of the bending force during silicide formation taken with the FLX-2908 and KMOS systems yielded similar results (Fig. 2). The sample in the FLX-2908 was annealed in a forming gas atmosphere. Because of the UHV environment, the temperature during the KMOS anneal may not be accurate. However, at ~ 350 °C, the KMOS measured a significantly larger tensile change than the FLX-2908. The shape of the second, larger compressive force also differed between the two measurements. These discrepancies suggested that the annealing ambient may have an effect on the stresses during silicide formation.

Figure 3 shows a plot of the stress data obtained from the FLX-2908 and the diffracted XRD intensity as a function of temperature and 2θ for a 100 nm Co/Si (100) sample annealed at 10 °C/min. In the 2θ range of 50-60° the following peaks were observed: HCP Co(002) and (101), Co$_2$Si(301), CoSi(210) and (211), and CoSi$_2$(220). Examining the upper plot, upon heating, a rapid relaxation of the deposition stress was evident from the decrease in the tensile stress between 210-250 °C. The HCP Co (002) was the only observed diffraction peak. This was consistent with previous experiments that reported film densification and recovery processes in Co films deposited on SiO$_2$[16]. Further heating led to an increase in compressive stress until ~ 375 °C where a slight change to a tensile direction occurred that could also be due to further Co grain growth or partial transformation to the FCC structure of Co[16]. A compressive force developed at ~400 °C commensurate with the formation and growth of the metal-rich silicide phase signified by the presence of the Co$_2$Si (301) peak in the XRD plot. The stress continued to

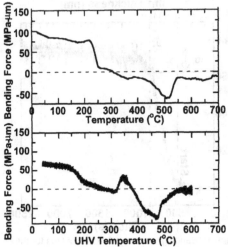

Figure 2: Comparison of bending force as a function of temperature obtained from the FLX-2908 (top) and the k-Space MOS (bottom).

increase in a compressive direction until the Co_2Si phase disappeared and the CoSi (210) and CoSi (211) peaks formed at ~500 °C. At this point, the stress reached a compressive maximum (510 °C). Further heating led to a relaxation in a tensile direction until ~560 °C where the $CoSi_2$ (220) peak rapidly appeared and the peaks from the CoSi phase began to disappear. Part of the reduction in $CoSi_2$ (220) could be a result of a change in the texture of the $CoSi_2$ phase from (220) to (111). Further heating to 700 °C did not produce any change in bending force. Upon cooling, the stress data retraced the heating curve until a deviation in the tensile direction began at ~550 °C. The average cooling rate was ~2 °C/min. The final values of the stress measured from the FLX-2908 runs are listed in table II. The second heating showed only a linear decrease in tensile stress, retraced the first cooling cycle and confirmed that phase formation was completed during the first temperature ramp.

Figure 4 shows a plot of the stress data obtained from the FLX-2908 and the diffracted XRD intensity as a function of temperature and 2θ for 35 nm Co films on poly-Si annealed at 10 °C/min. Following a slight increase in tensile stress beginning at ~150 °C a relaxation was observed beginning at 225 °C. The bending force increased in a nearly linear fashion in the compressive direction until ~375 °C. Although only HCP Co diffraction peaks were evident up to this temperature, we do not exclude the possibility of a partial HCP to FCC phase transformation in the Co. Unlike the formation of Co_2Si on Si

(100), at about 400 °C, the bending force remained constant and flat during the growth of this phase until it changed to a tensile direction beginning at ~460 °C. This temperature corresponded to onset of the CoSi (211) peak and occurred ~50 °C less than for the single crystal Si sample. The bending force continued to evolve in a tensile direction as the CoSi phase grew. Unlike the Si (100) sample, a net tensile force of ~50 MPa-μm was observed at 700 °C. The growth of the poly-Si grains was possibly the origin of this additional tensile stress [17]. The $CoSi_2$ (220) peak appeared at ~570 °C. The bending force was again constant after formation of the disilicide. Upon cooling, the stress curve

Figure 3: (Top) Plot of bending force vs. temperature for a 100 nm Co film on Si (100) annealed at 10 °C/min to 700 °C in FG. (Bottom) Plot of XRD intensity as a function of 2θ and temperature for the same film annealed in N_2. White indicates highest intensity.

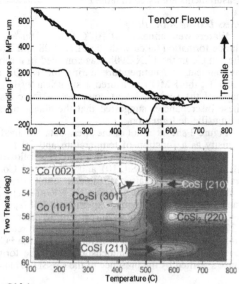

was retraced from 700° to 600° C after which a linear, tensile increase in the stress was observed. The room temperature stress was tensile with a final value of ~1.4 GPa.

DISCUSSION

One method to determine the intrinsic stresses resulting from the formation of a silicide phase is to compare the ratio of the resulting silicide volume (per molecule) to that of the reactants. This ratio results in values that are <1 for all Co-Si phases, thus predicting a tensile stress. The current experimental results indicated that this method is an incorrect description at least for the case of Co_2Si formation.

Another method of determining the stress change related to the sequential formation of the different silicides is to consider the primary diffusing species involved the reaction [4-5]. For instance, the Co_2Si/Si interface will move into Si by Co atom motion since Co is the primary diffusing species [18]. Using the appropriate volumes of Co_2Si and Si (i.e., 1 molecule of Co_2Si formed for every 1 Si atom), this reaction would cause a volume increase (V_{Co2Si}/V_{Si} = 1.63) leading to a compressive stress. This agrees qualitatively with the results from the FLX-2908 and KMOS annealing experiments. During the formation of CoSi, Si diffuses to the $Co_2Si/CoSi$ interface [18]. Again, the volume increases (V_{CoSi}/V_{Co2Si} = 1.33). In this case the comparison of the initial and final volumes of the reactants and products could describe the measured stress. However, we believe that the stress behavior at this temperature is dominated by relaxation mechanisms. Finally, the formation of $CoSi_2$ occurs via Co diffusion considering the temperature range of this study [19]. In this case, *two* reacting (moving) interfaces are present, $CoSi_2/CoSi$ = 0.89 and $CoSi_2/Si$ = 0.96. A decrease in volume occurs at both interfaces that would produce tensile stresses. Here, the two methods of determining stress would be in agreement. Since

Table II: Listing of stress values obtained from the FLX-2908 of Co films deposited on Si (100) and poly-Si after annealing to 700 °C in forming gas.

Sample	Subst.	Subst. Thick. (μm)	$CoSi_2$ Thick. (nm)	Stress (MPa)
100 nm Co	Si (100)	620	350	577
35 nm Co	Si (100)	620	123	581
35 nm Co	poly-Si	620	123	1173,1418

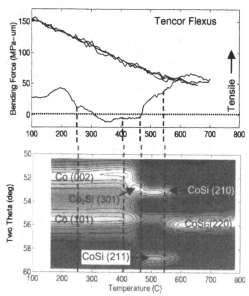

Figure 4: (Top) Plot of stress thickness vs. temperature for a 35 nm Co film on 150 nm poly Si annealed at 10 °C/min to 700 °C in FG. (Bottom) Plot of XRD intensity as a function of 2θ and temperature for the same film annealed in N_2. White indicates highest intensity.

Table III: Comparison of the volume change per atom between the silicide (V_{sil}) and either the volume of the reactants (V_{react}) or volume into which the reaction front moves (V_i).

Element	V_{react}	V_{sil}	V_{sil}/V_{react}	V_i	V_{sil}/V_i
Co	11.08	—	—		
Si	20.02	—	—		
Co_2Si	42.18	32.66	0.77 (tensile)	20.02 (Si)	1.63 (comp.)
CoSi	31.10	43.47	0.72 (tensile)	32.66 (Co_2Si)	1.33 (comp.)
$CoSi_2$	51.12	38.56	0.75 (tensile)	43.47 (CoSi)	0.89 (tensile)
				40.04 (Si)	0.96 (tensile)

no change in the bending force was observed during the formation of $CoSi_2$, the generated thermal and intrinsic stresses were accommodated by stress relief mechanisms such as creep or dislocation formation available at this annealing temperature. It should be noted that the relaxation of the compressive stress during the CoSi formation is most likely resulting from the same relaxation mechanisms. The volume calculations using the product and reactants or the dominant moving species are summarized in table III.

CONCLUSIONS

Stress changes of Co films on Si (100) and poly-Si have been investigated during deposition and isochronal annealing using both a KMOS and a KLA-Tencor Flexus FLX-2908 systems. The wafer curvature measurement during sputtering indicated that the as deposited stress was uniform through the thickness of the Co film. Both techniques indicated similar stress behavior during annealing and silicide formation. *In situ* synchrotron XRD data was correlated with FLX-2908 measurements acquired during annealing. This assisted in determining the identity of the silicide phases present at the different temperatures. Because of concurrent relaxation mechanisms, intrinsic stresses associated with silicide formation could not be predicted systematically using either reactant and product volumes or by considering the dominant moving species. Future experiments to decouple the operating stress mechanisms should clarify the origins of the experimentally observed stresses.

The authors thank Dr. J. Jordan-Sweet for beamline assistance and Drs. F.M. d'Heurle and I.C. Noyan at the IBM T.J. Watson Research Center for their insights. Beamline X20C is supported through DOE contract DE-AC02-76CH00016.

REFERENCES

[1] P.D. Agnello, S. Brodsky, E. Crabbe, E. Nowak, J. Lasky, and B. Davari, Electrochem. Soc. Proc., J1-Adv. in Rapid Thermal Processing May 2-6 1999.
[2] A. Steegen, I. DeWolf, and K. Maex, J. Appl. Phys., **86**, 4290 (1999).
[3] F.M. d'Heurle, Inter. Mater. Rev. **34**, 54 (1989).
[4] F.M. d'Heurle and O. Thomas, Defect and Diffusion Forum **129-130**, 137 (1996).
[5] S.L. Zhang, and F.M. d'Heurle, Thin Solid Films **213**, 34 (1992).
[6] L. Van den hove, Ph.D. Thesis, Katholicke Universiteit Leuven, B-3030 Leuven, Belgium (1988).
[7] A.R. Sitaram, J.C. Kalb, and S.P. Murarka, Mat. Res. Soc. Proc. **Vol. 188**, 67 (1990).
[8] J.A. Floro, E. Chason, and S.R. Lee, Mat. Res. Soc. Proc. **Vol. 406**, 491 (1996).
[9] k-Space Systems Inc., Ann Arbor MI.
[10] KLA-Tencor Inc., San Jose, CA.
[11] G.B. Stephenson, K.F. Ludwig, J.L. Jordan-Sweet, S. Brauer, J. Mainville, Y.S. Yang, and M. Sutton, Rev. Sci. Instrum. **60**, 1537 (1989).
[12] C. Cabral, Jr., L.A. Clevenger, J.M.E. Harper, R.A. Roy, K.L. Saenger, G.L. Miles, and R.W. Mann, Mat. Res. Soc. Proc., **Vol. 441**, (1997).
[13] S.-L. Zhang, C. Lavoie, C. Cabral, Jr., J.M.E. Harper, F.M. d'Heurle, and J. Jordan-Sweet, J. Appl. Phys. **85**, 2617 (1999).
[14] P. Koch, J. Phys. Condens. Matter. **6**, 9519 (1994).
[15] F. Spaepen, Acta Mater, **48**, 31 (2000).
[16] C. Cabral, Jr., K. Barmak, J. Gupta, L.A. Clevenger, B. Arcot, D.A. Smith, and J.M.E. Harper, J. Vac. Sci. Technol. A **11**, 1435 (1993).
[17] Q.Z. Hong, F. M. d'Heurle, J.M.E. Harper, and S.Q. Hong, Appl. Phys. Lett. **62**, 2637 (1993).
[18] G.J. van Gurp, W.F. van der Weg, and D. Sigurd, J. Appl. Phys. **49**, 4011 (1978).
[19] F.M. d'Heurle and C.S. Petersson, Thin Solid Films **128**, 283 (1985).

Mat. Res. Soc. Symp. Vol. 611 © 2000 Materials Research Society

Formation and Electrical Transport Properties of Nickel Silicide Synthesized by Metal Vapor Vacuum Arc Ion Implantation

X. W. Zhang, S. P. Wong, W. Y. Cheung and F. Zhang[1]
Department of Electronic Engineering, The Chinese University of Hong Kong, Shatin, N. T., Hong Kong, P. R. China

ABSTRACT

Nickel disilicide layers were prepared by nickel ion implantation into silicon substrates using a metal vapor vacuum arc ion source at various beam current densities to an ion dose of 6×10^{17} cm^{-2}. Characterization of the as-implanted and annealed samples was performed using Rutherford backscattering spectrometry, x-ray diffraction, electrical resistivity and Hall effect measurements. The temperature dependence of the sheet resistivity and the Hall mobility from 30 to 400 K showed peculiar peak and valley features varying from sample to sample. A two-band model was proposed to explain the observed electrical transport properties.

INTRODUCTION

Metal silicides have attracted widespread interest as contact and interconnect materials for silicon based microelectronic devices because of their low electrical resistivity, high chemical resistivity, good thermal stability, and compatibility to modern Si technology. Various techniques have been reported to synthesize nickel silicide [1-4], including ion beam synthesis (IBS) [4]. The advantages of the IBS method include the ease of control of the implant dose, the high reproducibility of the process, and the ability to directly form a buried layer. But the high dose requirement is a problem. However, the high dose problem is solved by using a metal vapour vacuum arc (MEVVA) ion source, which was first developed by Brown et al. in 1985 [5] and can provide very high beam current. In fact, there have been a number of reports in recent years that metal silicide layers with good crystal quality and good electrical properties can be formed by MEVVA implantation [4, 6, 7]. In this paper, we shall report the formation and electrical transport properties of NiSi$_2$ layers synthesized by MEVVA implantation.

EXPERIMENTAL DETAILS

NiSi$_2$ layers were prepared on p-Si (100) substrates of 10-20 Ω cm by nickel implantation using a MEVVA ion source. The MEVVA source was operated in a pulse mode. The implantation was performed at an extraction voltage of 65 kV to an ion dose of 6×10^{17} cm^{-2} at various mean ion current densities from 13 to 52 μA/cm^2. During implantation, the vacuum level of the target chamber was 8×10^{-4} Pa. Rapid thermal annealing was performed in an Ar ambient at 700 °C for 40 s. Glancing incidence x-ray diffraction (XRD) was performed to identify the phases of the synthesized silicides. Rutherford backscattering spectrometry (RBS) experiments were performed at a 170° scattering angle with a 2 MeV ^{4}He^{2+} ion beam. The sheet

[1] Visiting from Shanghai Institute of Metallurgy, Chinese Academy of Sciences, Shanghai, China
Present address: University Heidelberg, Physikalisch-Chemisches Institut/Radiochemie, Germany

resistivity and Hall effect measurements were carried out using the van der Pauw's method from 30 to 400 K.

RESULTS AND DISCUSSION

The glancing angle XRD spectra of three samples implanted at several beam current densities to a dose of 6×10^{17} cm^{-2} are shown in Fig. 1. The implant beam current densities are 13, 26, and 52 μA/cm^2 for sample A, B and C, respectively. It can be seen that a unique NiSi$_2$ phase has been formed in these samples. For sample A of the lowest beam current density, only some weak diffraction peaks are observed. For sample B and C prepared with higher beam current densities, all diffraction peaks of NiSi$_2$ are observed and the intensity of the peaks is strengthened, signaling a better crystallinity.

Fig. 2 shows a typical RBS spectrum of a sample implanted with a mean beam current density of 26 μA/cm^2 to a dose of 6×10^{17} cm^{-2}. A clear step just behind the Si leading edge in the spectrum confirms the formation of NiSi$_2$ on Si. By fitting the RBS spectrum using the RUMP simulation program [8], the atomic composition ratio of Si:Ni and the thickness of the NiSi$_2$ layer was estimated to be 2.15:1 and 56 nm, respectively. The obtained composition ratio and the thickness of the NiSi$_2$ layers from their RBS spectra as well as the sample preparation conditions are listed in Table I. It can be seen that the composition ratio is close to that of a stoichiometric NiSi$_2$ (2:1) for all these samples. The XRD and RBS results indicate that NiSi$_2$ layers have been directly formed in these as-implanted samples. This is consistent with the findings of earlier reports on nickel silicide formation by MEVVA implantation [4]. In conventional IBS, high temperature annealing is normally required to form the NiSi$_2$ phase. However, for IBS using MEVVA implantation, because of the pulse mode operation, the

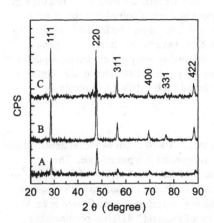 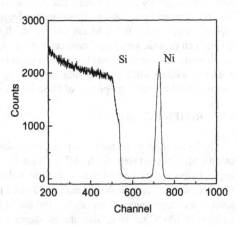

Figure 1. *Glancing angle XRD spectra of the samples implanted at a beam current density of (A) 13μA/cm^2, (B) 26μA/cm^2 and (C) 52μA/cm^2 to a dose of 6×10^{17} cm^{-2}.*

Figure 2. *RBS spectrum of the sample implanted with a mean beam current density of 26 μA/cm^2 to a dose of 6×10^{17} cm^{-2}.*

Table I. Preparation conditions and characterization results of the NiSi$_2$ layers formed by Ni implantation into Si (100) at an extraction voltage of 65 kV to a dose of 6×10^{17} cm^{-2}

Sample	Beam current density ($\mu A/cm^2$)	Thermal annealing (RTA 700°C/40s)	Thickness (nm)	Si:Ni	Sheet resistivity at 300K (Ω/Sq)	Resistivity at 300K ($\mu\Omega$ cm)
A	13	N	54	2.33	55.2	298
B	26	N	56	2.15	13.5	75.6
C	52	N	42	1.88	29.8	125
D	26	Y	57	2.08	12.9	73.5

instantaneous beam current density and hence the input power density during implantation is very high. This can lead to substantial ion beam heating effect and/or ion beam induced crystallization effect which are believed to be among the possible mechanisms responsible for the direct formation of the NiSi$_2$ phase.

Results of the Hall effect measurements indicate that the conduction type of the silicide layers is p-type. Listed in Table I are also the estimated values of the resistivity at 300 K from the respective data of the sheet resistivity and the layer thickness. The resistivity of sample B, implanted at a beam current density of 26 $\mu A/cm^2$, is about 75 $\mu\Omega$ cm, somewhat larger than but comparable to the reported resistivity of NiSi$_2$ [9-11]. However, for samples A and C, implanted at a higher and a lower current density, the obtained resistivity values are even higher. This might be attributed to a geometry effect. The NiSi$_2$ layers are not perfect though electrically they form a continuous conducting layer. There might be Si islands/channels embedded in the layer so that the effective conduction path length is larger and the effective cross-section smaller. After annealing, the resistivity at 300 K are generally smaller than their as-implanted counterparts.

The temperature dependence of the electrical transport properties of these layers from 30 to 400 K exhibit very interesting and unexpected characteristics. For NiSi$_2$ thin films formed by conventional methods, the electrical resistivity ρ shows metallic behavior and $\rho(T)$ obeys the Matthiessen's rule [11]. But as shown in Fig. 3a and 3b, the temperature dependence of the sheet resistivity and the Hall mobility for these samples exhibit peculiar behavior that has not been observed before for NiSi$_2$. The sheet resistivity showed peak and valley features, and with the detailed shape varying from sample to sample. The Hall mobility showed a big peak with the peak temperature and peak value dependent on sample. It is also found that despite of the fact that the three as-implanted samples showed quite different temperature dependence, they showed very similar temperature dependence after annealing, with only some shift in the absolute magnitude. This difference in magnitude can again be attributed to the geometry effect mentioned above. Therefore for simplicity, in Fig. 3, we have only shown the results for one of the annealed samples.

After some work, it was found that the seemingly very complicated electrical transport properties of these samples could well be described by a simple phenomenological two-band model. We suppose that there are effectively two valence bands, labeled by 1 and 2, of the implanted p-type NiSi$_2$. Band 1 has a lower energy and a smaller mobility μ_1 and band 2 has a higher energy and a larger mobility μ_2. The band 1 in this model can be regarded as the

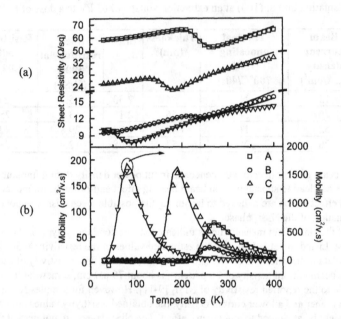

Figure 3. Temperature dependence of (a) the sheet resistivity and (b) the Hall mobility for the samples implanted at a beam current density of (A) 13 $\mu A/cm^2$, (B) 26 $\mu A/cm^2$, (C) 52 $\mu A/cm^2$ to a dose of 6×10^{17} cm^{-2} and (D) the sample B after rapid thermal annealing in Ar at 700 ^0C for 40 s. Experimental data are represented by the symbols, and the lines are the fitting results using the two-band model.

conventional valence band in normal NiSi$_2$. A possible mechanism responsible for the emergence of the band 2 could be attributed to the stress existing in the implanted NiSi$_2$ layers, hence leading to some stress-induced band structure changes. Let the energy separation between the two bands be $\Delta E = kT_e$, where k is the Boltzmann's constant. Assume also that the total number of carriers, n, simply does not depend on the temperature, but the number of carriers in band 1, n_1, and that in band 2, n_2, can vary with temperature with their ratio obeying the Boltzmann distribution, i.e.

$$n_1(T) + n_2(T) = n = \text{constant} \tag{1a}$$

$$\frac{n_2}{n_1} = K \times \exp(-\frac{T_e}{T}) \tag{1b}$$

where K represents the ratio of the density of states of the two bands. In general, K is temperature dependent but if it is a slow varying function of temperature, for simplicity we can just take it as a constant.

Now it becomes quite plausible that the peculiar temperature dependence of the transport properties could come from the changes in the number of carriers with temperature in the two bands. We further assume that the temperature dependent mobilities in the two bands are given by the following equations:

$$\frac{1}{\mu_1} = \frac{1}{\mu_{10}} + (\frac{T}{T_1})^{3/2} \tag{2a}$$

$$\frac{1}{\mu_2} = (\frac{T_a}{T})^r + \frac{B}{\exp(T_b/T) - 1} \tag{2b}$$

where μ_{10}, T_1, T_a, T_b, r and B are fitting parameters. The first and second terms at the right hand side of Eq. (2a) and (2b) correspond to the contributions from the dominant scattering mechanisms at low and high temperatures, respectively. The choices of the forms of these terms are discussed in the following. We fixed the power of T to 3/2 in the second term of the right hand side of (2a) to represent the contribution of acoustic phonon scattering. For the first term in (2a), it could generally take a similar form as that in (2b), to represent the dominant scattering mechanisms at low temperatures. In the present case, however, we found that it suffices to take it as a constant, hence reducing the number of fitting parameters by one. On the other hand, the second term in (2b) could more simply take a form in some positive power of T. Nevertheless we found that the form we chose resembled that of the contribution from optical phonon scattering and in fact it led to better fitting of the experimental data. We did prepare thin $NiSi_2$ layers by deposition of thin Ni layers on Si plus thermal treatment, and measure their electrical transport properties. These 'normal' $NiSi_2$ layers showed normal metallic behaviour. The temperature dependent μ_1, obtained from Eq. (2a) is in good agreement with the experimental results of the 'normal' $NiSi_2$ samples.

Using equations (1) and (2) of the two-band model, it is straightforward to show that the sheet resistivity R_s and Hall mobility μ_H can be expressed as

$$R_s = \frac{\rho}{t} = \frac{1}{t(n_1 q\mu_1 + n_2 q\mu_2)} = \frac{1 + \frac{n_2}{n_1}}{N(\mu_1 + \mu_2 \frac{n_2}{n_1})} \tag{3}$$

$$\mu_H = \frac{\mu_1^2 n_1 + \mu_2^2 n_2}{\mu_1 n_1 + \mu_2 n_2} = \frac{\mu_1^2 + \mu_2^2 \frac{n_2}{n_1}}{\mu_1 + \mu_2 \frac{n_2}{n_1}} \tag{4}$$

where q is the elementary charge, t is the effective layer thickness, and $N = nqt$ is in effect another fitting parameter, taking care of the geometry effect mentioned earlier, to fit the absolute magnitude of the resistivity data.

Using Eqs. (1) to (4), the temperature dependence of the sheet resistivity and the Hall mobility of the $NiSi_2$ layers can be fitted. The fitting results were plotted as lines in Fig. 3, which is in good agreement with the experimental data. The parameter r is determined to be 3/2 for all annealed samples, which indicates that the first term in Eq. (2b) comes from the contribution of impurity scattering for the annealed samples. But for the as-implanted samples, the value of r is sample dependent and takes a much larger value of typically 15. This suggests that the transport process and the scattering mechanisms in the as-implanted samples are much more complicated. We speculate that this is associated with the stress states induced by MEVVA implantation in the as-implanted samples. There is normally some release of stress after thermal annealing. This may also correlates with the fact that despite of the very different temperature dependence before annealing, the temperature dependence of the transport properties after annealing becomes similar. One may also speculate that it could be related to the earlier mentioned geometry effect. The layer contains small Si islands/channels which could lead to significant carrier scattering especially at lower temperatures. This speculation can probably be checked by future transmission electron microscopy study.

CONCLUSIONS

In summary, we have formed NiSi$_2$ thin layers directly by Ni implantation into Si using a MEVVA ion source. The temperature dependence of the sheet resistivity and the Hall mobility of the as-implanted NiSi$_2$ layers show very interesting and unexpected characteristics with peak and valley features not reported before in the literature. A simple phenomenological two-band model was proposed to understand these transport properties. The fitting results according to the two-band model are in good agreement with the experimental data.

ACKNOWLEDGMENTS

This work is partially supported by the Research Grants Council of Hong Kong SAR (Ref. No. CUHK4405/99E). Two of us (WYC and FZ) are also supported by a Direct Grant for Research from the Faculty of Engineering of CUHK.

REFERENCES

1. Z. Tan, F. Namavar, J. I. Budnick, F. H. Sanchez, A. Fasihuddin, S. M. Heald, C. E. Bouldin and J. C. Woicik, *Phys. Rev.*, **B46**, 4077 (1992).
2. J. Y. Yew, L. J. Chen and K. Nakamura, *Appl. Phys. Lett.*, **69**, 999 (1996).
3. D. X. Xu, S. R. Das, C. J. Peters and L. E. Erickson, *Thin Solid Films*, **326**, 143 (1998).
4. K. Y. Gao, B. X. Liu, *Appl. Phys.*, **A68**, 333 (1999).
5. I.G. Brown, J. E. Gavin and R. A. MacGill, *Appl. Phys. Lett.*, **47**, 358 (1985).
6. Q. Peng, S.P. Wong, I.H. Wilson, N. Wang, K.K. Fung, *Thin Solid Films* **270**, 573 (1995).
7. Q. Peng and S.P. Wong, *Mat. Res. Soc. Symp. Proc.* **402**, 487 (1996).
8. L.R. Doolittle, *Nucl. Instrum. Meth.* **B5**, 344 (1985).
9. J. C. Hensel, R. T. Tung J. M. Poate and F. C. Unterwald, *Appl. Phys. Lett.*, **44**, 913 (1984).
10. S. P. Murarka, *Intermetallics*, **3**, 173 (1995).
11. F. Nava, K. N. Tu, O. Thomas, J. P. Senateur, R. Madar, A. Borghesi, G. Guizzetti, U. Gottlieb, O. Laborde and O. Bisi, *Mater. Sci. Rept.*, **9**, 141 (1993).

Poster Session

Mat. Res. Soc. Symp. Vol. 611 © 2000 Materials Research Society

The Physical and Electrical Properties of Polycrystalline $Si_{1-x}Ge_x$ as a Gate Electrode Material for ULSI CMOS Structures

Sung-Kwan Kang[a] Dae-Hong Ko[a], Tae-Hang Ahn[b], Moon-Sik Joo[b], In-Seok Yeo[b],

Sung-Jin Whoang[c], Doo-Young Yang[c], Chul-Joo Whang[c,] Hoo-Jeong Lee[d]

[a]Dept. of Ceramic Engineering, Yonsei University, Seoul, Korea

[b]Hyundai Electronics Industries Co. Ltd., Memory Research & Development Division, Kyungki-do, Korea

[c]Ju-Sung Co. Ltd., Kyungki-do, Korea

[d]Stanford University, Stanford, California 93405, USA

ABSTRACT

Poly $Si_{1-x}Ge_x$ films have been suggested as a promising alternative to the currently employed poly-Si gate electrode for CMOS technology due to lower resistivity, less boron penetration, and less gate depletion effect than those of poly Si gates. We investigated the formation of poly $Si_{1-x}Ge_x$ films grown by UHV CVD using Si_2H_6 and GeH_4 gases, and studied their microstructures as well as their electrical characteristics. The Ge content of the $Si_{1-x}Ge_x$ films increased linearly with the flux of the GeH_4 gas up to x=0.3, and saturated above x=0.45. The deposition rate of the poly $Si_{1-x}Ge_x$ films increased linearly with the flux of the GeH_4 gas up to x=0.1, above which it is slightly changed. The resistivity of the $Si_{1-x}Ge_x$ films decreased as the Ge content increased, and was about one half of that of poly-Si films at the Ge content of 45%. The C-V measurements of the MOSCAP structures with poly $Si_{1-x}Ge_x$ gates demonstrated that the flat band voltage of the poly $Si_{1-x}Ge_x$ films was lower than that of poly-Si films by 0.2V.

INTRODUCTION

As the size of the CMOS device scales down, conventionally used polycide or salicide gate stacks with poly Si films show serious issues of gate depletion effect, boron penetration and so on [1-3]. Though mid-gap metal gates have been suggested as an alternative, the process is incompatible with conventional Si integration processes[4-8]. Recently, poly $Si_{1-x}Ge_x$ films have been suggested as a promising alternative to the currently employed poly Si gate electrode for CMOS technology due to the low resistivity, variable workfunction, and compatibility with Si processes[9-12].

In this paper, we investigated the microstructures and electrical characteristics of poly $Si_{1-x}Ge_x$ films using Si_2H_6 and GeH_4 as a deposition source gas in a UHV CVD system. We found that poly $Si_{1-x}Ge_x$ films could be used as an alternative to conventional poly Si films.

EXPERIMENT

Poly $Si_{1-x}Ge_x$ films with 0%, 15%, 30% and 45% Ge content were deposited using UHV CVD (EUREKA

2000, Ju-Sung Co. Ltd.) on the 8" silicon wafers on which a 1000Å thick thermal SiO_2 layers were grown. The deposition temperature ranged from $550°C$ to $625°C$. The base pressure prior to the deposition process was below $1 \times 10^{-7} torr$, and the process pressure ranged from 1mtorr to 5mtorr with 100% Si_2H_6 and 10%, 100% GeH_4 as source gases. In order to reduce incubation time for the formation of poly $Si_{1-x}Ge_x$ films and enhance the adhesion of poly $Si_{1-x}Ge_x$ on SiO_2[13], 100Å thick poly Si films were deposited as a seed layer, on which the poly $Si_{1-x}Ge_x$ films were deposited.

After depositing 1000Å thick poly $Si_{1-x}Ge_x$ with the Ge content of 0%, 15%, 30%, and 45%, 100Å or 1500Å thick poly Si capping layers were deposited in order to prevent Ge out-diffusion from poly $Si_{1-x}Ge_x$ films and to prevent the formation of unstable GeO during dopant activation process. After deposition process, boron implantation with the dose of $1 \times 10^{15}/cm^2$, $3 \times 10^{15}/cm^2$ was carried out. Implantation energies were 5keV for 100Å poly $Si/1000 \text{Å}$ poly $Si_{1-x}Ge_x$, and 20keV for 1500Å poly $Si/1000 \text{Å}$ poly $Si_{1-x}Ge_x$ structures. Following the implantation process, the films were annealed for 30 sec at the temperatures between $600°C$ and $900°C$ in order to activate the dopants. Deposition rate, Ge content, surface roughness, and crystal structure were analyzed using SEM, XRD, RBS, and XPS. The dopant distribution and resistivity of poly $Si_{1-x}Ge_x$ films were measured by SIMS and four point probe measurement. The C-V and I-V characteristics of the MOSCAP structures with poly $Si_{1-x}Ge_x$ gates were measured for 50Å thick SiO_2 as a gate oxide layer.

RESULTS

1. Deposition rate and Ge content of Poly $Si_{1-x}Ge_x$

Fig. 1(a) shows the changes of the deposition rate of poly $Si_{1-x}Ge_x$ films deposited by UHVCVD process using 100% Si_2H_6 and 100% GeH_4 as source gases. The deposition rate increased linearly with GeH_4 gas flux up to 10sccm, and then saturated at $200 \text{Å}/min$, $250 \text{Å}/min$, and $280 \text{Å}/min$ at the deposition temperature of $575°C$, $600°C$, and $625°C$, respectively.

Fig. 1(b) shows Ge content of poly Si1-xGex as a function of GeH4 flux when Si2H6 gas flux was fixed at 20 sccm. Up to 30%, Ge content increased linearly with GeH_4 gas flux, above which Ge content slightly increased and saturated at about 45%.

2. Dopant activation and Ge redistribution in poly $Si_{1-x}Ge_x$ films upon annealing

The sheet resistance of the boron implanted poly $Si_{1-x}Ge_x$ films was measured by four point probe before and after annealing treatment. Fig. 2 shows that sheet resistance values of poly $Si_{1-x}Ge_x$ decrease with the increase of the Ge content in the films at all annealing temperatures. The decrease of the sheet resistance is due to the higher dopant activation rate and larger grain size than those of poly Si films

Fig. 3(a) is the SIMS depth profile, which shows the boron distribution in the poly $Si_{0.55}Ge_{0.45}$ films after annealing at $750°C$, $850°C$, and $950°C$ for 30 sec. As the annealing temperature increased, the boron penetration into the poly $Si_{1-x}Ge_x$ films increased. After annealing at $950°C$, the boron concentration was almost evenly distributed in the films. The boron profiles in poly $Si_{1-x}Ge_x$ films with different Ge contents after $950°C$ annealing are shown in Fig. 3(b). The diffusion of boron in poly Si is shown to be faster than that in poly $Si_{1-x}Ge_x$, and consequently the diffusion of boron decreases with the increase of the Ge content in poly $Si_{1-x}Ge_x$ films. The lower diffusivity of boron in poly $Si_{1-x}Ge_x$ is consistent with the previous results by Salm et al.[14].

In addition, Ge redistribution occurs in due to the diffusion of Ge into the capping Si layers, resulting in the reduction of the Ge content in poly $Si_{1-x}Ge_x$ films. Fig. 4 shows that Ge content of poly $Si_{0.7}Ge_{0.3}$ is reduced to x=0.15 after annealing at 1100°C for 30 sec.

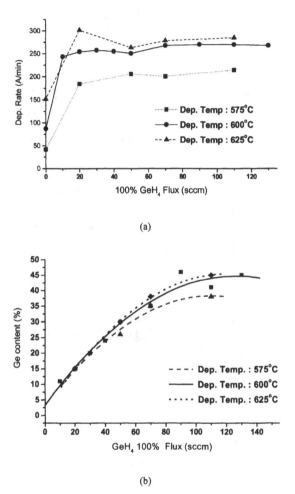

(a)

(b)

Figure 1. (a) Deposition rate and (b) Ge content as a function of GeH_4 gas flux at fixed Si_2H_6 flux of 20sccm at the deposition temperature of 575 °C, 600 °C, and 625 °C

3. C-V Characteristics of poly $Si_{1-x}Ge$ films

The C-V characteristics of the MOSCAP structures with poly $Si_{1-x}Ge_x$ gates are shown in Fig. 5. The flat band voltages of the poly $Si_{1-x}Ge_x$ films are lower than that of poly Si films by 0.15-0.2V. The flat band voltage shift is attributed to the decrease of the work function of poly $Si_{1-x}Ge_x$ films compared with that of the poly Si films.

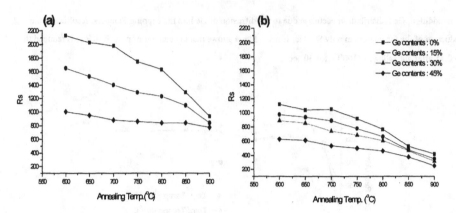

Figure 2. *Sheet resistance values of poly Si$_{1-x}$Ge$_x$ films implanted with boron at 5keV with a dose of (a) 1 ×10^{15}/ cm^2 and (b) 3 ×10^{15}/cm^2*

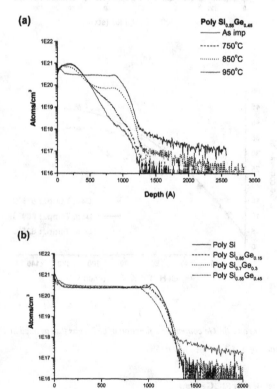

Figure 3. *(a) SIMS depth profiles of boron in poly Si$_{0.55}$Ge$_{0.45}$ films after annealing at various temperatures for 30 sec in N$_2$ ambients, (b) SIMS depth profile of boron in poly Si$_{1-x}$Ge$_x$(x=0, 0.15, 0.3, 0.45) after annealing at 950 °C for 30 sec. in N$_2$ ambients*

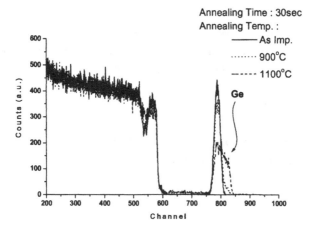

Figure 4. RBS spectra of poly Si 1500 Å/poly $Si_{0.7}Ge_{0.3}$ 1000 Å/ seed/ SiO_2 1000 Å/Si after annealing at various temperature for 30 sec in the N_2 gas ambient

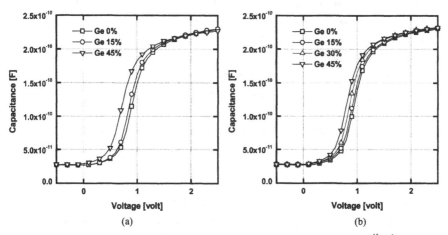

Figure 5. C-V characteristics of PMOS capacitors implanted with boron at dose of (a) $1 \times 10^{15}/cm^2$ and (b) $3 \times 10^{15}/cm^2$

CONCLUSIONS

We investigated the formation of poly $Si_{1-x}Ge_x$ films using UHV CVD with Si_2H_6 and GeH_4 as source gases. The Ge content of the $Si_{1-x}Ge_x$ films increased linearly with the flux of the GeH_4 gas up to x=0.3, and saturated above x=0.45. The deposition rate of the $Si_{1-x}Ge_x$ films increased linearly with the flux of the GeH_4 gas up to x=0.1, above which it is slightly changed. The resistivity of the $Si_{1-x}Ge_x$ films decreased as the Ge content increased, and was about one half of that of poly-Si films at the Ge content of 45%. The C-V measurements of the MOSCAP structures with poly $Si_{1-x}Ge_x$ gates demonstrated that the flat band voltage of the poly $Si_{1-x}Ge_x$ films was lower than that of poly-Si films by 0.2V.

ACKNOWLEDGEMENTS

This research is funded by ministry of science & technology and ministry of commerce, industry, and energy of KOREA, which we gratefully appreciate.

REFERENCES

1. G. J. Hu and R.H. Bruce, IEEE Trans Electron Devices, **vol. 32**, No. 3, pp 584 (1985)

2. M. Hao, D. Nayak and R. Rakkhit, IEEE Electron Device Lett., **vol. 18**, No. 5, pp 215 (1997)

3. B. Yu, D.H. Ju, N.Kepler, T.J. King and C. Hu, Symp. VLSI., pp105 (1997)

4. N. Yamamoto, S. Iwaa, N. kobayashi and T.Tedra, Proc 15th SSDM, pp. 217 (1983)

5. K.Ohnishi, N. Yamamoto, T. Uchino, Y. Hanaoka, R. Tsuchiya, Y. Nonaka, Y. Tanabe, T. Umezawa, N. Fukuda, S. Mitani and T. Shiba, IEDM Tech. Dig., pp 397 (1998)

6. H. Shimada, Y. Hirano, T. Ushiki and T. Ohmi, IEDM Tech. Dig., pp 881 (1995)

7. Y. Okazaki, T. Kobayashi, H. Inokawa, S. Nakayama, M. Miyake, T. Morimoto and Y. Yamamoto, IEEE Trans. Electron Devices, **vol. 42**, No. 9, pp. 1583 (1995)

8. H. Koike, Y. Unno, F. Maysuoka and M. Kakumu, IEEE Trans. Electron Devices, **vol. 44**, No. 9, pp. 1460 (1997)

9. T.J. King, J.R. Pfiester, J.D. Shott, J.P. McVittie and K.C. Saraswat, IEDM Tech. Dig., pp 253 (1990)

10. Y.V. Ponomarev, C. Salm, J. Schmitz, P.H. Woerlee, P.A. Stolk and D. J. Gravesteijn, IEDM Tech. Dig., pp 829 (1997)

11. V. Z-Q Li, M.R. Mirabedini, R.T.Kuehn, J.J. Wortman and M. C. öztürk, IEDM Tech. Dig., pp 833 (1997)

12. W.C. Lee, T.J.King and C. Hu, IEEE Electron Device Lett., **vol. 20**, No. 1, pp 9 (1999)

13. H.C. Lin, C.Y. Chang, W.H.Chen, W.C.Tsai, T.C.Chang and H. Y. Lin, J. Electrochem. Soc., **vol 141**, No. 9, pp.2559 (1994)

14. C. Salm, D.T. van Veen, D.J. Gravesteijn, J. Hollerman and P.H. Woerlee, J. Electrochem. Soc., **vol 144**, No. 10, pp3665. (1997)

Mat. Res. Soc. Symp. Vol. 611 © 2000 Materials Research Society

A Low-Cost BiCMOS Process with Metal Gates

H.W. van Zeijl, and L.K. Nanver
Laboratory of ECTM, DIMES, Delft University of Technology
P.O.Box 5053, 2600 GB, Delft, The Netherlands
Phone: +31 15 2784949 Fax: +31 15 2622163 E-mail: henkz@dimes.tudelft.nl

ABSTRACT

A low-complexity and low-cost double-metal BiCMOS process is proposed with only 13 mask steps. By decoupling the source-drain thermal budget from gate-stack formation, metal gates are realizable.

INTRODUCTION

Scaling BiCMOS technology will in general increase the process complexity and reduce the process flexibility. Together with the increasing costs of an advanced process line, research related to such technology can become forbiddingly expensive. The research capabilities are also compromised by high process complexity, which reduces the compatibility with alternative processing modules, for example for the integration of specialized electronic components and smart sensors [1]. This work proposes a novel device architecture that enables submicron device dimensions but maintains a moderate process complexity that moreover allows an attractive degree of flexibility. Significant process simplification is achieved because the process flow mainly contains steps that are common to both the bipolar and the CMOS device formation. A key point is the decoupling of the thermal budget of source-drain formation from the gate-stack formation while still maintaining a self-aligned device structure. This gives an extra degree of freedom in the selection of both the gate dielectric and the gate electrode material, thus making this structure very suitable as a research vehicle for the study of alternative gate stacks. The total process presented here is a 13 mask double-metal BiCMOS process that can be fabricated with conventional process equipment. It features high frequency BJT's and metal gate MOS devices.

PROCESS DESCRIPTION

The BiCMOS process flow is show schematically in figure 1. First a 0.9 µm epitaxial silicon layer is grown on a p-type (100) substrate with implanted buried n^+ layers. The dopant concentration in the epitaxial layer is 10^{16} cm^{-3} and serves as collector for the NPN and as n-well. Next a high-energy, high-dose phosphorus implant is performed to form the collector plug. A high-energy, high-dose boron implant contacts the substrate and provides junction isolation for the bipolar transistors. The collector plug and p-isolation also contact the n-well and p-well, respectively, and are used as channel stopper. The p-well is formed by a high-energy, low-dose boron implantation that gives a dopant concentration of 10^{17} cm^{-3}. A 200 nm low-stress silicon nitride film (SiN$_x$, x=0.94) is deposited directly on the silicon by LPCVD (field nitride). In the active device areas the field nitride is plasma etched followed by a thermal oxidation of 20 nm of silicon dioxide. Next a 200 nm thick polysilicon film is deposited. This film is heavily doped by ion implantation: boron is implanted above the PMOS and the NPN's and phosphorus above the NMOS. A 200 nm SiN$_x$ film is then deposited on the polysilicon. This second nitride and polysilicon are plasma etched selectively to the oxide and the underlying nitride to open the emitter and gate windows. By wet chemical etching the thermal oxide under the doped polysilicon is under-etched to form a 0.3 μm long cavity. This cavity is filled by a second undoped polysilicon deposition. Thermal oxidation and subsequent wet chemical etching removes the excess polysilicon. In this way, polysilicon filled cavities around the emitter and gate windows are formed and they act as a diffusion path for the dopants in the polysilicon above source, drain and extrinsic base.

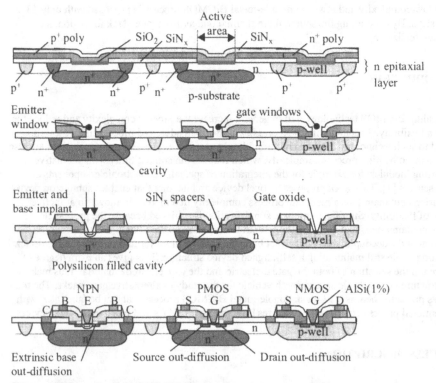

Figure 1 *Schematic of the BiCMOS process flow*

In the gate and emitter windows, spacers of SiN$_x$ are formed. Before the spacer nitride is deposited, the native oxide is removed by dip-etching. The silicon exposed in the gate and emitter windows is oxidized and this 30 nm thick oxide is removed by wet chemical etching. Since nitride spacers are used this wet etch does not change the dimensions of the emitter/gate windows. Once again a 30 nm thick gate oxide is grown by thermal oxidation. In the NPN transistors the gate oxide is removed using a non-critical mask that protects the gate oxide in the MOS devices. The same mask is used to implant the arsenic emitter and the boron intrinsic base in the emitter window. Thus the intrinsic base is implanted after the emitter window etch and spacer formation whereby the NPN current gain is not sensitive to silicon loss during postprocessing such as oxidation or etching. Contact holes are etched through the silicon nitride to the poly- and mono-crystalline silicon. All implanted dopants are activated using a single thermal anneal of 30 min at 950 °C in argon ambient. Before metallization with AlSi(1%) a dip-etch in 0.5% HF is performed and the gate oxide is reduced to a thickness of 15 nm. After the first metallization the process is completed with a conventional second metallization. The total process requires only 13 photomasks.

DEVICE CHARACTERISTICS AND DISCUSSION

In figure 2 a Gummel plot is shown of a typical NPN transistor with an emitter size of $0.5 \times 40 \ \mu m^2$. The non-ideal base current observed at low emitter-base forward bias is low and typical of NPN's with low-stress nitride spacers [2]. The corresponding output characteristic, the cut-off frequency and the maximum oscillation frequency are shown in figures 3 and 4.

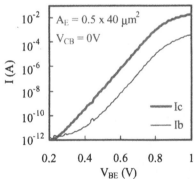

Figure 2 *Gummel plot of a NPN transistor.*

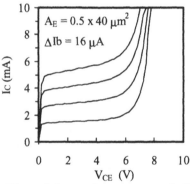

Figure 3 *Output characteristics of a NPN transistor.*

In this process the extrinsic base is self-aligned to the emitter and intrinsic base. The polysilicon filled cavities around the emitter act as a diffusion path for dopants from the highly-doped polysilicon to the monsilicon. This out-diffusion provides the baselink to the intrinsic base and determines the electrical characteristics of the emitter sidewalls. If voids or other structural irregularities were present in the filled cavities, the out-diffusion would also be irregular and at worst it would be blocked. This would result in a spread in current gain and possibly emitter-collector shorts at the emitter perimeter.

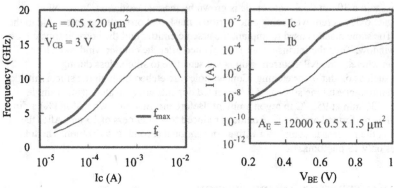

Figure 4 f_t and f_{max} versus collector current.

Figure 5 Gummel plot of an NPN transistor array with 12000 emitters.

On large transistor arrays with 12000 emitters of 0.5 x 1.5 μm^2 (figure 5) no such shorts were found and current gain fluctuations in individual devices were also not observed. This indicates that the filled cavities form a reliable diffusion path.

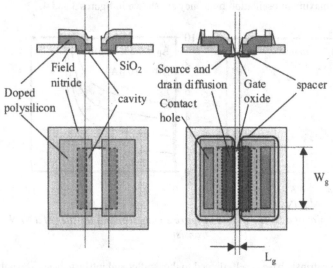

Figure 6 Top view of the MOS device design.

A top view of the MOS device design is shown in figure 6. In contrast to common device isolation technologies like LOCOS or shallow trench isolation, a low-stress silicon nitride film restricts the source and drain dopant diffusions from the polysilicon near the gate edges. For the NMOS, parasitic inversion layer currents under the field nitride must be suppressed by extra p-type doping of the surface. This can be achieved by locating the p-isolation implantation at a distance of 1 μm from the gate edges. This eliminates the inversion currents without extra mask steps and without increasing the threshold voltage of long devices. The subthreshold characteristics for MOS devices with a gate length of 0.5 μm are show in figures 7 and 8. The threshold voltage for the NMOS and PMOS is 0.63 V and 0.94 V, respectively, and the threshold voltage versus gate length is shown in figure 11. The output characteristics of the PMOS and NMOS are show in figures 9 and 10, respectively, and the device parameters are summarized in Table I.

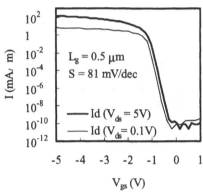

Figure 7 PMOS subthreshold characteristics.

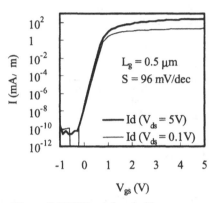

Figure 8 NMOS subthreshold characteristics.

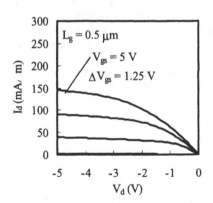

Figure 9 NMOS output characteristics.

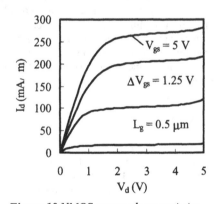

Figure 10 NMOS output characteristics.

Table I Device parameters

	NPN	
Emitter area	20 x 0.5	μm^2
$f_{tmax}(V_{cb} = 3V)$	13	GHz
$f_{max}(V_{cb} = 3V)$	20	GHz
Va	25	V
H_{fe}	95	
BV_{ceo}	8	V
R sheet extr. base poly	169	$\Omega/\square$
	NMOS	
$I_{ON\ Vd = 5V}$	275	$\mu A/\mu m$
V_t	0.63	V
R sheet sourc-drain poly	49	$\Omega/\square$
	PMOS	
$I_{ON\ Vd = 5V}$	150	$\mu A/\mu m$
V_t	0.94	V
R sheet extr. base poly	169	$\Omega/\square$

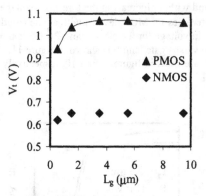

Figure 11 Threshold voltage versus gate lengths.

CONCLUSIONS

In the process presented here, basically only three processing steps are limited to one application in the formation of the three transistor types: the p-well implantation for the NMOS, the n-doping of the NMOS source-drain poly and the implantation of the NPN emitter and intrinsic base. All other steps have multiple purposes. The use of implanted emitters, metal gates and inside nitride spacers also assures a low complexity process where conventional thermal processing and 1 μm lithography suffices for achieving submicron device dimensions. The poly filled cavities that connect the polysilicon extrinsic base/source/drain to the monosilicon provide submicron contacts to these regions, resulting in for example low collector-base capacitance and high f_{max}. The device measurements indicate that the filling of this cavity is a reliable processing step. All in all this is a robust process with a low number of mask steps and high flexibility.

REFERENCES

1   G. C. M. Meijer, *Concepts and focus point for intelligent sensor systems*, Sensor and Actuators A, 41-42 1994 p. 183-191
2   H. W. van Zeijl and L.K. Nanver, *Characterization of low-stress LPCVD silicon nitride in high frequency BJT's with self-aligned metallization.* ISICT Beijing 1998 , p 98-101

Mat. Res. Soc. Symp. Vol. 611 © 2000 Materials Research Society

EFFECTS OF NITRIDATION BY N$_2$O OR NO ON THE ELECTRICAL PROPERTIES OF THIN GATE OR TUNNEL OXIDES

C.Gerardi, T.Rossetti, M.Melanotte
STMicroelectronics, Central Research & Development, Catania Technology Center, Stradale Primosole 50, 95121 Catania, Italy
S.Lombardo, I.Crupi
Istituto Nazionale di Metodologie e Tecnologie per la Microelettronica (IMETEM), Consiglio Nazionale delle Ricerche, Stradale Primosole 50, 95121 Catania, Italy

ABSTRACT

We have studied the effects of nitridation on the leakage current of thin (7-8 nm) gate or tunnel oxides. A polarity dependence of the tunneling current has been found this behavior is related to the presence of a thin silicon oxynitride layer at the SiO$_2$/Si-substrate interface. The oxynitride layer lowers the tunneling current when electrons are injected from the interface where the oxynitride is located (substrate injection). The current flowing across the oxide when electrons are injected from the opposite interface (gate injection) is not influenced by the oxynitride. The increase of nitrogen concentration leads to a decrease of the tunneling current for substrate electron injection.

INTRODUCTION

Nitridation of thin oxides obtained by annealing with nitrous oxide (N$_2$O) and nitric oxide (NO) is used in ULSI technology because it improves the dielectric reliability [1-3]. These processes are attractive due to their simplicity and because they are hydrogen-free thus avoiding the problems related to H that generates electron traps.

The nitridation processes obtained by using N$_2$O and NO are similar, in fact in both cases the species responsible for the nitridation is NO. However, during N$_2$O nitridation only a small amount of NO is produced by the dissociation of N$_2$O at high temperature, in addition the oxygen by-product of such dissociation continues the oxidation increasing the final oxide thickness [4]. Direct nitridation in NO is more effective since it does not involve N$_2$O dissociation and therefore it can be exploited at lower temperature [5-6].

In this work we have studied the effects of nitridation on the current flowing across the oxide in the Fowler-Nordheim (F-N) regime finding a correlation with the amount of the nitrogen and its position in the oxide. The experiments have been performed on thin oxides grown by wet oxidation and subjected to annealing in NO. Several conditions have been investigated by varying the annealing time and the flux of NO in order to achieve different N concentrations.

EXPERIMENTAL

Oxides with a nominal thickness of 7-8 nm were obtained by thermal steam oxidation of p(100) Si substrates. The oxides were designed to be utilized as thin tunnel dielectrics for nonvolatile memory applications.

After oxidation the wafers were annealed in NO at a temperature of 850°C with different annealing times and NO fluxes. For comparison one of the oxides (control) was annealed only in N_2 with the same thermal budget.

Secondary ion mass spectrometry (SIMS) analyses were carried out by using a 2 keV Cs^+ primary ion beam and detecting NCs^+, OCs^+ and $SiCs^+$ molecular ions.

Current-Voltage (I-V) and Capacitance-Voltage (C-V) measurements were performed on MOS capacitors with an area of 1×10^{-3} cm^2 , fabricated with a standard LOCOS isolation process.

I-V measurements under accumulation condition were performed by negatively biasing the gate. For I-V measurements under MOS strong inversion, the gate and substrate were positively biased and an additional current supply was connected to an n-type well which, by forward biasing the substrate-well p-n junction, acts as supply of electrons for the formation of the MOS inversion layer.

RESULTS

Figure 1 shows the nitrogen, oxygen and silicon SIMS profiles relative to the samples subjected to NO annealing. One can observe that nitrogen is piled-up at the SiO_2/Si-substrate, exhibiting a distribution with an FWHM of ~ 2.5 nm. The nitrogen content has been quantified by calculating the area under the nitrogen SIMS signal in fact, due to resolution limitation, the true nitrogen profile is distorted and the peak of the SIMS signal is not indicative of the true concentration. SIMS shows that the increase of annealing time and NO flux leads to an increment of nitrogen incorporation at the SiO_2/Si-substrate interface.

The increase of nitrogen content in the oxide causes a shift in the flat band voltage and induces some distortions in the quasi-static C-V curves as it can be observed in Fig.2.

The leakage currents measured across the oxides under both accumulation and inversion conditions are reported in Fig.3. The I-V characteristics on the right side and on the left side of the figure are obtained under accumulation and inversion, respectively. The continuous lines plotted in Fig.3 are the theoretical F-N curves calculated for both substrate (inversion) and gate (accumulation) injection. The theoretical I-V curves under accumulation and inversion have been obtained by strictly assuming a F-N tunneling with an oxide barrier of 3.15 eV, an electron effective mass of 0.5 m_0 and neglecting the image charge terms [9].

Starting from the control oxide we have fitted the leakage currents both in accumulation and inversion with F-N curves using the electric field in the oxide and flat-band voltage as parameters.

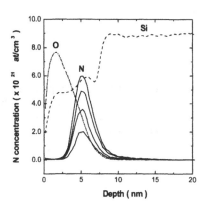

Fig.1: *SIMS profiles of the nitrided oxides. The N profiles are related to the following concentrations: a) 6.4×10^{14} at/cm², b) 11.0×10^{14} at/cm², c) 16.0×10^{14} at/cm² and d) 20.0×10^{14} at/cm². The profiles of O and Si are not in scale.*

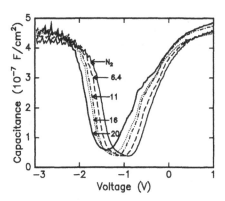

Fig.2: *Quasi-Static C-V curves of the investigated oxides, including the control oxide (labelled with N_2). The arrows indicate the N content $\times 10^{14}$ at/cm².*

The voltage drop in the oxide V_{ox} is given by:

$$V_{ox} = E_{ox}T_{ox} = V_G - V_{FB} - \boxtimes - V_P \qquad (1)$$

where E_{ox} is the electric field across the oxide, T_{ox} is the oxide thickness V_{FB} is the flat band voltage, $\boxtimes$ is the Si surface potential and V_P the voltage drop in the poly.

As it can be observed in Fig.3, a good fitting can be achieved both in accumulation and inversion for the oxide that has not been subjected to nitridation. The oxide thickness (T_{ox}) is derived from (1) by fitting the theoretical F-N curves in accumulation and inversion to the experimental I-V curves.

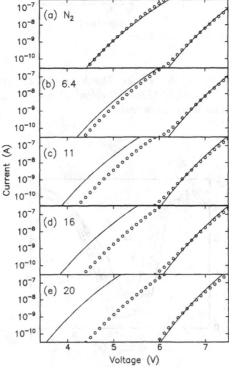

Fig.3: *I-V data (open circles) measured under both gate and substrate injection. The lines are the theoretical F-N curves.*

Using the same procedure we have tried to fit the I-V curves of nitrided oxides in the F-N regime, the fit has been performed by following the trend of flat band shifts and the oxide thickness derived from C-V measurements in accumulation condition. Nevertheless for the nitrided oxides the fitting procedure works only under accumulation, i.e. for electron injection from the gate. Under inversion condition, when electrons are injected from the interface where the oxynitride layer is localized, the I-V characteristics cannot be fitted without changing other parameters such as the tunneling barrier height. In addition, the deviations from the F-N behavior under inversion conditions increase with the concentration of incorporated nitrogen.

It is known that the dielectric constant of an oxynitride increases and the tunneling barrier height decreases almost linearly when the nitrogen relative concentration in the oxynitride is augmented [12]. The increase of nitrogen content leads to an increment of the dielectric constant of the thin oxynitride layer at the SiO_2/Si-substrate interface, raising the electron tunneling distance. A polarity dependence of the F-N tunneling current, i.e., different behavior in the leakage current according to whether substrate or gate injection is considered, has been observed in stack Si_3N_4/SiO_2 or $SiON$/SiO_2 dielectrics and it has been ascribed to the asymmetry of the barrier shape [13-14].

The nitrided oxides studied in this work can be considered as stack dielectrics consisting of an SiO_2 layer and a thin SiON or even Si_3N_4 layer for prolonged annealing in NO. The asymmetric behavior of the leakage current shown by the nitrided oxides is explained by considering that, due to the presence of the oxynitride layer, the resulting tunneling distance is higher for substrate injection.

CONCLUSION

We show that nitridation leads to the formation of a nitrogen-rich layer at the SiO_2/Si-substrate interface, this layer causes a lowering of the tunneling current across the dielectric for electrons injected from this interface. The nitrided oxides can be considered as SiO_2/SiON stacks in which the SiON approaches the Si_3N_4 stoichiometry as the nitrogen content is augmented. This leads to an increment in the dielectric constant of the nitrided layer, therefore the tunneling distance results higher when electrons are injected from the nitrided SiO_2/Si-substrate interface.

REFERENCES

1. D.J. Di Maria, E.Cartier, and D.Arnold, J. Appl. Phys. **73**, 3367 (1993).
2. K.F. Schuegraf, and C.Hu, Semiconductor Sci. Technol. **9**, 989 (1994).
3. T. Hori, "Gate Dielectrics and MOS ULSIs, Principles, Technologies and Applications, Springer - Verlag, Berlin - Heidelberg (1997) p.209.
4. Y. Okada, Ph. J. Tobin, K. G. Reid, R.I. Hedge, B. Maiti and S. A. Ajuria, IEEE Trans. Electron Devices **41**, 1608 (1994).
5. E.P. Gusev, H.C. Lu, T. Gustafsson, E. Garfunkel, M.L. Green and D. Brasen, J. Appl. Phys. **82**, 896 (1997).

6. C.Gerardi, R.Zonca, B.Crivelli and M.Alessandri, J. Electrochem Soc.**146**, 3059 (1999).
7. B.Y Kim, L.K.Han, D.Wristers, J. Fulford, and D.L. Kwong, in "The Physics and Chemistry of SiO2 and the Si - SiO2 Interface - 3, H.Z.Massoud, E.H. Poindexter, C.R. Helms Eds., Proc. Vol. 96 - 1, The Electrochemical Society, NJ (1996) p.772.
8. M.Bhat, L.K.Han, D.Wristers, J. Yan, D.L. Kwong and J. Fulford, Appl. Phys. Lett. **66**, 1225 (1995).
9. Y. Nissan-Cohen, J. Shappir and D. Frohman - Bentchowsky, J. App. Phys. **54**, 5793 (1983).
10. T. Hori, "Gate Dielectrics and MOS ULSIs, Principles, Technologies and Applications, Springer - Verlag, Berlin - Heidelberg (1997) p. 23.
11. S.M.Sze, "Physics of Semiconductor Devices", second edition, J. Wiley & Sons, NY, (1981).
12. Xin Guo, T.P. Ma, IEEE Electron Device Letters **19**, 207 (1998).
13. Y.Shi, T.P.Ma, Sharad Prasad, IEEE Trans. Electron Devices **45**, 2355 (1998).
14. Shi, X. Wang, T.P.Ma, IEEE Trans. Electron Devices **46**, 362 (1999).

Mat. Res. Soc. Symp. Vol. 611 © 2000 Materials Research Society

Hot Wall Isothermal RTP for Gate Oxide Growth and Nitridation

Allan Laser, Christopher Ratliff, Jack Yao, Jeff Bailey, Jean-Claude Passefort, Eric Vaughan, Larry Page
Silicon Valley Group, Thermal Systems Division, Scotts Valley, CA 95066, U.S.A.

ABSTRACT

A new system that incorporates many benefits of large batch furnaces (high quality films, growth of wet and dry oxides, chlorine capability, and low cost) into a single wafer processing module has been developed at SVG Thermal Systems. The problems associated with wafer temperature measurement and control in traditional lamp based RTP systems are avoided by utilizing a hot wall isothermal processing chamber. Unique fixturing is used to minimize thermal stress on the wafer during ramping. High quality gate oxides ranging in thickness from 20Å to 40Å have been grown in this system using both wet and dry oxidation ambients, with and without chlorine. Thin oxides grown in dry oxygen had 1-sigma uniformities in the range of 0.72-0.95%, while oxides grown in oxygen/HCl (1-3%) had uniformities of 0.80%. Steam grown oxides demonstrated growth rates of 100Å/min at 900°C and uniformities of 0.62%. Dry oxides annealed in NO and N_2O had peak nitrogen incorporation levels ranging from 0.5 to 5.1 atomic percent depending on anneal ambient, temperature and time.

INTRODUCTION

Two trends in the manufacturing of advanced silicon devices have resulted in a shift of oxidation processes towards single wafer thermal processors. The first such trend comes about from the fact that wafer batch size has been reduced in the manufacturing process in both logic and ASIC fabs, where many different device types are often processed in small batch sizes. The small batch is desired to improve cycle time. In some of these cases, the cost of ownership (CoO) advantage long enjoyed by batch furnaces compared to single wafer processing no longer prevails. This trend will continue to accelerate as wafer production moves to 300 mm.

The second trend is related to the necessity of preparing multi-processed gate dielectrics for the formation of advanced gate dielectric stacks. No longer is a simple wet or dry gate oxidation required, but in many cases, the gate oxide must be nitrided to enhance its resistance to boron penetration [1] and to increase the device reliability [2]. Nitridation is often accomplished by multi-step processing that requires the use of several different reactive gas species [3]. The smaller volume of a single wafer tools allows such reactive gases to be quickly replaced in a sequential manner resulting in improved process repeatability.

Most single wafer oxidation tools are lamp heated systems. Some of these systems have difficulty growing high quality oxides in steam. As a result of the materials of construction (e.g., stainless steel) some lamp-based tools cannot grow the chlorinated oxides that are required to achieve dielectric properties comparable to oxides grown in batch furnaces. These tools also suffer from temperature measurement variability related to variations in the wafer emissivity. In this paper we describe a new single wafer rapid thermal processing tool that utilizes a hot wall isothermal chamber that maintains the advantages of batch furnaces, but does not suffer from reduced throughput for small batch sizes nor sensitivity to wafer emissivity variations. Results on the wafer show excellent process repeatability and uniformity, and slip free wafers. Since the chamber is constructed from fused silica (quartz) it can readily tolerate chlorinated oxidations in both dry oxygen and steam ambients. In this paper we describe the major features of this new tool and process results for chlorinated dry and steam oxidations. We also present the results of nitrogen incorporation using NO and N_2O annealing.

EXPERIMENTAL

Hardware configuration

The main objectives when designing this single wafer hot wall RTP system were to retain the positive aspects of furnace processing while improving the controllability over process chamber gases and improving heat flux uniformity to the wafer surface.

The process chamber consists of a quartz process tube surrounded by a multizone heating element. The combination of symmetric, strategically placed heating zones, along with a silicon carbide thermal diffusion plate allows the system to achieve excellent thermal uniformity in the upper portion of the process tube. The wafer is heated by elevating it from a lower chill chamber to an upper hot chamber. A shutter system thermally isolates the hot and cold chambers. The shutters open and close only as the wafer passes from chamber to chamber. By isolating the hot and cold chambers, better thermal uniformity in the process area is achieved and the energy efficiency of the system is improved. It should be mentioned that the energy consumption of a hot wall system is much less than that of a lamp based system, and peak power requirements are also greatly reduced.

Uniform wafer temperature during ramp and steady state is made possible by transporting the wafer in a quartz wafer holder with custom features incorporated to minimize edge to center temperature differences. The system is also capable of wafer rotation that aids in thermal uniformity as well as improving the gas distribution across the wafer.

The wafer is exposed to a cross flow of process gases only in the upper portion of the process tube. The cross flow is a laminar plug type flow which allows quick transition from one gas species to another. The injector plenum is designed to give uniform flow distribution across the wafer surface. The uniformity has been substantiated by flow visualization.

One of the key advantages of this system is that all of the parts that are exposed to process gases are made of either quartz or silicon carbide. This provides the capability of performing HCl assisted steam cleans for improved removal of metallic contaminants and HCl assisted oxidation.

Oxidation tests

All oxidations were performed on 200mm p-type <100> particle grade test wafers with the wafer rotating at 6 rpm. No precleans were performed prior to oxidation. Dry oxides were grown in 100% O_2. Wet oxides were grown in a variety of steam concentrations where the steam was generated using a pyrogenic external torch. Chlorinated oxides were grown in both dry and wet ambients with 1% to 3% HCl added during oxidation. Nitrided oxides were first grown in 100% O_2 and then followed by nitridation in 100% NO or N_2O. Processing took place at atmospheric pressure or slightly negative pressure (5 to 10 Torr below ambient). Film thickness uniformity was evaluated using a Rudolph Caliber 300 Ellipsometer with a 49-pt, 3mm edge exclusion pattern. Nitrogen incorporation was characterized by Secondary Ion Mass Spectrometry (SIMS) depth profiling. Wafers were examined for slip using the Magic Mirror [4] technique and for surface dislocations using the Secco etchant [5].

RESULTS

Thermal performance

Temperature data taken using a 200mm thermocouple instrumented wafer indicates that for a process temperature of 1050°C, the average edge to center temperature difference falls below 0.25°C before the oxidation process starts. Instantaneous ramp rates exceed 65°C/second during the ramp. No evidence of slip was observed in any of the wafers processed for these experiments.

Process performance

Table I lists various dry oxidation temperatures and times. The average oxide thickness ranges from 23Å to 36Å. Figure 1 shows an example of a wafer map with contour lines at 0.1Å intervals for a dry oxide demonstrating a thickness of 22.5Å with 0.80% 1-sigma uniformity.

Table I. Dry oxidation conditions and uniformity results

Temp	Oxidation Time	O₂ Flow	Average Thickness	Uniformity 1-sigma
900°C	30 sec	20 slpm	22.5Å	0.80%
900°C	35 sec	20 slpm	24.1Å	0.95%
965°C	60 sec	20 slpm	35.0Å	0.81%
1000°C	20 sec	20 slpm	32.4Å	0.72%
1000°C	30 sec	20 slpm	35.9Å	0.75%

One of the features of this new tool is the ability to grow oxides in a chlorinated ambient (by adding HCl gas) during the oxidation step. Table II shows the oxide results for a 20 second dry oxide grown at 1000°C with chlorine concentrations varying from 1% to 3%. The average thickness increases with increasing chlorine concentration as expected but within wafer uniformity stays constant at 0.81% 1-sigma.

Table II. Chlorinated dry oxide conditions and uniformity results

Temp	Oxidation Time	O₂ Flow	HCl Flow	Average Thickness	Uniformity 1-sigma
1000°C	20 sec	6.5 slpm	65 sccm	33.2Å	0.81%
1000°C	20 sec	6.5 slpm	135 sccm	34.1Å	0.81%
1000°C	20 sec	6.5 slpm	200 sccm	35.5Å	0.80%

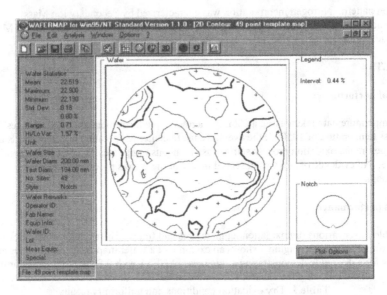

Figure 1. Uniformity map of 900°C, 30 sec, dry oxide

Table III lists a variety of wet oxidation tests where the average thickness ranged from 25Å to 40Å. Nitrogen was used to dilute the steam during oxidation in order to slow the growth rate for most of these tests. Figure 2 shows a map of within wafer uniformity with contour lines at 0.1Å intervals for a 26.3Å wet oxide which had a uniformity value of 0.48% 1-sigma.

Table III. Wet oxidation conditions and uniformity results

Temp	Oxidation Time	H₂ Flow	O₂ Flow	N₂ Flow	Average Thickness	Uniformity 1-sigma
850°C	60 sec	7.2 slpm	4.8 slpm	6 slpm	28.9Å	0.79%
900°C	15 sec	3.0 slpm	8.0 slpm	10 slpm	25.2Å	0.62%
900°C	20 sec	4.0 slpm	3.2 slpm	10 slpm	26.3Å	0.48%
900°C	40 sec	4.0 slpm	8.0 slpm	none	40.4Å	0.65%
950°C	15 sec	3.0 slpm	8.0 slpm	none	37.8Å	0.98%
1000°C	15 sec	2.0 slpm	8.0 slpm	15 slpm	36.6Å	0.69%

The use of chlorine during oxidation is also possible for wet oxides. Table IV lists the results obtained by varying the HCl concentration from 1% to 3% during a typical wet oxide process. Increasing the HCl concentration does not impact average thickness significantly and within wafer uniformity remains stable.

Table IV. Chlorinated wet oxide conditions and uniformity results

Temp	Oxidation Time	H₂ Flow	O₂ Flow	HCl Flow	Average Thickness	Uniformity 1-sigma
900°C	30 sec	2.6 slpm	5.2 slpm	65 sccm	32.1Å	0.74%
900°C	30 sec	2.6 slpm	5.2 slpm	135 sccm	32.7Å	0.79%
900°C	30 sec	2.6 slpm	5.2 slpm	200 sccm	32.7Å	0.74%

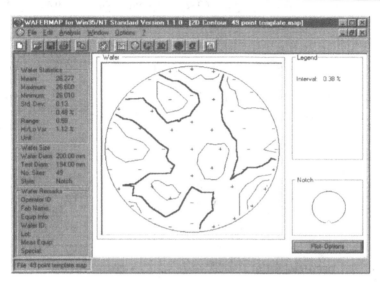

Figure 2. Uniformity map of 900°C, 20 sec, 4.0/3.2/10.0 slpm H₂/O₂/N₂ wet oxide

The system is capable of growing or annealing oxides in NO or N₂O in order to create oxynitride gate dielectrics. Table V and VI list the conditions and results for various nitrided films. For oxides annealed in NO or N₂O the initial dry oxide thickness was 32Å at 1000°C and 34Å at 1050°C. Thickness uniformities after nitridation are typically below 1% 1-sigma and peak nitrogen incorporation ranges from 0.5 to 5.1 atomic percent. Figure 3 shows the SIMS profile for a NO and N₂O annealed oxide.

Table V. N₂O oxynitride conditions, uniformity results, and nitrogen incorporation levels

Temp	Ox Time	O₂ Flow	N₂ Flow	Anneal Time	N₂O Flow	Avg Tox	Unif 1-sigma	Peak Nitrogen
1000°C	60 sec				5 slpm	32.0Å	0.76%	1.4 at.%
1000°C	20 sec	10 slpm		30 sec	5 slpm	38.4Å	0.41%	0.5 at.%
1000°C	20 sec	10 slpm		60 sec	5 slpm	41.1Å	0.42%	1.0 at.%
1050°C	60 sec				5 slpm	39.5Å	0.70%	2.4 at.%
1050°C	10 sec	5 slpm	15 slpm	30 sec	5 slpm	42.3Å	0.60%	1.1 at.%
1050°C	10 sec	5 slpm	15 slpm	60 sec	5 slpm	45.6Å	0.53%	2.1 at.%

Table VI. NO oxynitride conditions, uniformity results, and nitrogen incorporation levels

Temp	Ox Time	O_2 Flow	N_2 Flow	Anneal Time	NO Flow	Avg Tox	Unif 1-sigma	Peak Nitrogen
1000°C	20 sec	10 slpm		30 sec	5 slpm	37.6Å	0.89%	1.4 at.%
1000°C	20 sec	10 slpm		60 sec	5 slpm	39.3Å	1.07%	2.5 at.%
1050°C	10 sec	5 slpm	15 slpm	30 sec	5 slpm	37.6Å	0.69%	3.3 at.%
1050°C	10 sec	5 slpm	15 slpm	60 sec	5 slpm	40.9Å	0.44%	5.1 at.%

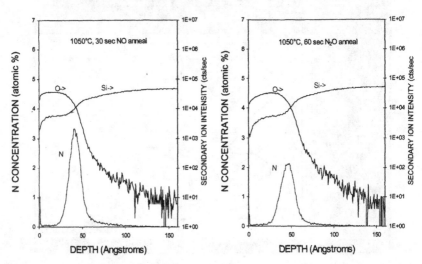

Figure 3. Nitrogen profiles for dry oxides annealed in NO and N_2O at 1050°C

CONCLUSIONS

SVG Thermal Systems has developed a new single wafer hot wall isothermal RTP that incorporates many of the benefits of batch furnaces. The system is capable of producing wet and dry oxides, with and without chlorine, all with excellent uniformities. Oxidations or anneals in NO or N_2O can be performed to produce nitrided gate oxides.

ACKNOWLEDGEMENTS

The authors would like to thank the SVG teams involved in the development of this new RTP (single wafer furnace) system. Thanks are also due to Rich Tauber for helpful discussions.

REFERENCES

1. H.G. Pomp, et al., IEDM Tech Dig. p. 463 (1993).
2. H. Hwang, et al., Applied Phys. Lett. **57**, 3 (1990).
3. J. Kuehne, et al., Rapid Thermal and Integrated Processing VI, (Materials Research Society Symposium Proceedings) **470**, 387 (1997).
4. Trademark of Hologenix, 15301 Connector Lane, Huntington Beach, CA.
5. F. Secco d'Aragona, J. Electrochem. Soc. **119**, 948 (1972).

Mat. Res. Soc. Symp. Vol. 611 © 2000 Materials Research Society

Controlling CoSi₂ nucleation : the effect of entropy of mixing

C. Detavernier[*], R.L. Van Meirhaeghe[*], K. Maex[+0], F.Cardon[*]

[*] Laboratorium voor Kristallografie en Studie van de Vaste Stof, Universiteit Gent,
 Krijgslaan 281/S1, B-9000 Gent, Belgium.
[+] IMEC, Kapeldreef 75, B-3001 Leuven, Belgium.
[0] also at E.E. Dept, K.U. Leuven, B-3001 Leuven, Belgium.

ABSTRACT

It is generally known that nucleation effects strongly influence the CoSi to CoSi₂ phase transition. According to classical nucleation theory, the small difference in Gibbs free energy between the CoSi and CoSi₂ phase is responsible for the nucleation barrier. Adding elements that are soluble in CoSi and insoluble in CoSi₂ will influence the entropy of mixing, and thus change ΔG. In this way, the height of the nucleation barrier may be controlled.
By depositing Fe or Ge (respectively replacing Co and Si in the CoSi lattice) in between the Co and the Si substrate, we were able to increase the nucleation barrier. In the presence of Ni, the nucleation barrier is lowered, and low-resistive disilicide is formed at lower temperatures.

INTRODUCTION

It is known from literature that when a film of Co is deposited onto a Si substrate, annealing results in sequential growth of Co_2Si, $CoSi$ and $CoSi_2$. It has been reported that the formation of $CoSi_2$ is nucleation controlled [1,2]. The activation energy for nucleation is given by [3]

$$\Delta G^* \approx \frac{(\Delta \sigma)^3}{(\Delta G)^2} = \frac{(\Delta \sigma)^3}{(\Delta H - T \Delta S)^2} \tag{1}$$

with $\Delta \sigma$ the difference in interfacial energy caused by the formation of the nucleus and ΔG the change in free energy. From eqn. 1, it is clear that nucleation phenomena will only be important if ΔG is small. In the case of $CoSi_2$, the small difference in free energy ΔH between the CoSi and $CoSi_2$ is responsible for the nucleation barrier. However, ΔG also contains an entropy term ΔS, which is usually neglected for solid state reactions. In this work, it will be shown that one can control the nucleation temperature of $CoSi_2$ by changing this ΔS term. This can be accomplished by adding alloying elements that are soluble in CoSi or $CoSi_2$. We were able to control the nucleation temperature in the range 400-800°C.

EXPERIMENT

The substrates (p-type Si(100), $N_a = 10^{13}$-10^{14} cm^{-3}) were cleaned using standard RCA cleaning, followed by a HF dip. Layers of Fe, Ge, Ni and Co were deposited by e-beam evaporation in a vacuum of 10^{-6} mbar. To study the silicidation reaction, isochronal annealing (30 seconds) was done at various temperatures. Annealing was carried out in a Rapid Thermal Processing (RTP) system, in N_2 ambient. The samples were analyzed using a four point probe for sheet resistance measurements, X-Ray Diffraction (XRD, CuKα radiation) for phase identification and X-ray Photoelectron Spectroscopy (XPS) to determine elemental depth distribution. Because of the large difference in resistivity between CoSi and CoSi$_2$, sheet resistance measurements provide a fast means of obtaining information about phase formation. In this paper, because of the limited amount of space, we will only present the sheet resistance measurements as 'evidence' for phase formation. However, more detailed phase identification was done by XRD.

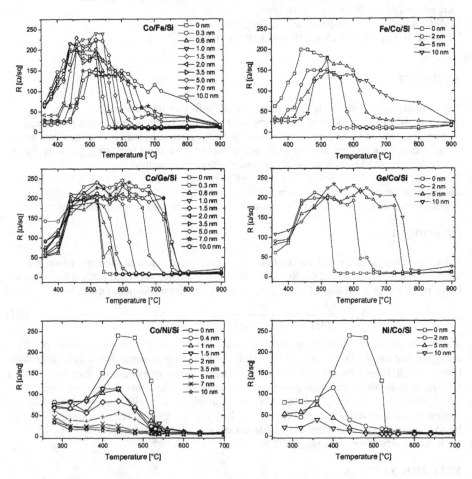

Figure 1 : Sheet resistance versus annealing temperature for Fe/Co/Si, Co/Fe/Si, Co/Ge/Si, Ge/Co/Si, Co/Ni/Si and Ni/Co/Si structures. The Co layer thickness is 10 nm for all samples, while the capping or interlayer thickness is indicated in the legend of each graph.

RESULTS

Nucleation temperature

From the sheet resistance versus annealing temperature plot (figure 1), it is clear that in the presence of Fe (both interlayer and capping layer) the formation of the low resistive $CoSi_2$ phase is delayed. For Fe interlayers that are thinner than 1 nm, the delay in nucleation temperature is very small (less than 10°C). However, for thicker interlayers, there is a gradual increase of the $CoSi_2$ nucleation temperature. For Fe capping layers, a similar behavior was observed as for interlayers. Moreover, once nucleated, the growth of the $CoSi_2$ layer is slowed down. This may be explained by the expelled Fe that precipitates at the $CoSi_2$ grain boundaries, thus slowing down diffusion. For a Co(10nm)/Fe(10nm)/Si system, $FeSi_2$ precipitates are observed by XRD.

To study the effect of Ge on the nucleation of $CoSi_2$, Co/Ge/Si (interlayer) or Ge/Co/Si (capping layer) structures were deposited on Si(100) and after annealing, phase formation was studied. Both for Ge interlayers and capping layers, a similar behavior was observed as for Fe. From the sheet resistance versus annealing temperature graph (figure 1), one observes that the low resistive disilicide phase forms at increasingly higher temperature for increasing amounts of Ge. In case of Ge, the delay in nucleation temperature saturates at about 750°C for a Ge interlayer thickness larger than 3.5 nm. The $CoSi_2$ nuclei may be observed by phase contrast microscopy (figure 2).

Figure 2 : Phase contrast microscopy image of a Co(10nm)/Ge(3.5nm)/Si sample annealed at 725°C for 30s.

In a third experiment, bilayers of Co/Ni/Si (interlayer) or Ni/Co/Si (capping layer) were deposited on Si(100). In the case of Ni, the sheet resistance (figure 1) indicates that the low resistive disilicide phase forms at lower temperature as compared to the standard Co/Si reaction, in contrast to the increase that was observed for Fe and Ge. For increasing thickness of the Ni interlayer, a gradual decrease of the nucleation temperature is observed. It should be noted that in this case, the decrease in sheet resistance does not necessarily imply the formation of $CoSi_2$, since NiSi also has a low resistivity (ρ = 18 and 10.5 $\mu\Omega$cm for $CoSi_2$ and NiSi, as compared to 110, 147 and 34 $\mu\Omega$cm for Co_2Si, CoSi and $NiSi_2$). However, the formation of disilicide was verified by XRD measurements (not shown here). Previously, d'Heurle et al. also observed a decrease in $CoSi_2$ formation temperature in the presence of Ni [4].

Miscibility

To check the miscibility of FeSi and CoSi, the position of the (210) CoSi peak was measured as a function of Fe layer thickness and annealing temperature (figure 3). It is found that the $Co_{1-x}Fe_xSi$ formed at 440°C obeys Vegard's law : the lattice parameter of the $Co_{1-x}Fe_xSi$ compound changes linearly as a function of the Fe content. However, the shift in peak position decreases for increasing annealing temperature, indicating that Fe is slowly expelled from the $Co_{1-x}Fe_xSi$.

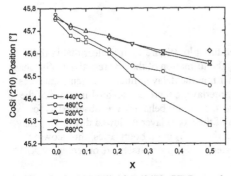

Figure 3 : Monosilicide (210) XRD peak position as a function of $x = d_{Fe}/(d_{Fe}+d_{Co})$ and annealing temperature.

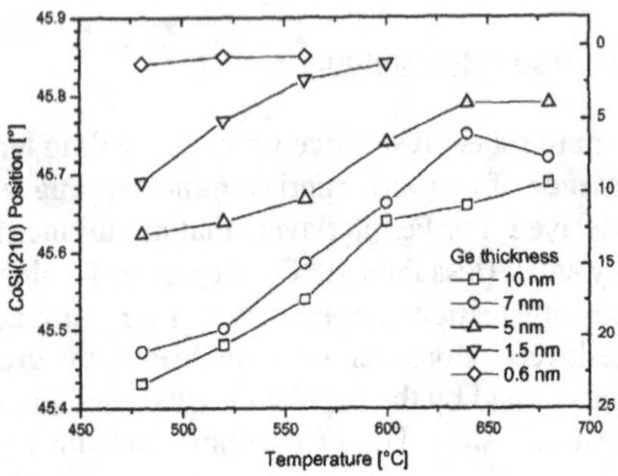

Figure 4 : Monosilicide (210) XRD peak position as a function of Ge thickness and annealing temperature.

Similar measurements were done for the Co/Ge/Si system. Wald et al. [5] studied the bulk ternary Co-Si-Ge system and reported that up to 67% of Ge is soluble in CoSi. Furthermore, they observed that the lattice parameter of the $CoSi_{1-x}Ge_x$ compound changes linearly as a function of the Ge content. For our samples, the position of the (210) monosilicide peak was measured as a function of interlayer thickness and annealing temperature. Especially for low annealing temperatures, a shift in the peak position is observed. However, the shift is smaller than may be expected if all the Ge would be dissolved into the CoSi. From XPS, it was found that for increasing annealing temperature, an increasing amount of Ge is piling up at the interface, forming a SiGe solid solution.

The expulsion of Fe and Ge from the mixed CoSi can be explained by thermodynamics. CoSi, FeSi and CoGe have cubic crystalline structures with lattice constans of 4.43, 4.49 and 4.68Å, respectively and are therefore expected to be miscible. However, the heat of formation of CoSi (-100.5 kJ/mole) is higher than that of FeSi (-80.4 kJ/mole) and CoGe (-34.2 kJ/mole). In the presence of a large supply of Si, Co-enriched $Co_{1-x}Fe_xSi$ and $CoSi_{1-x}Ge_x$ will be formed. For Ge, the expulsion is stronger than for Fe, because of the larger difference in heat of formation.

DISCUSSION

In the standard reaction to form $CoSi_2$, $CoSi + Si \rightarrow CoSi_2$, ΔS is usually neglected, a common procedure for solid-state reactions. Suppose however that some element A is completely soluble in CoSi and insoluble in $CoSi_2$. In this case, the reaction is given by

$$Co_{1-x}A_xSi + (1-2x)Si \longrightarrow (1-x)CoSi_2 + xA \qquad (2)$$

and there will be a difference in mixing entropy between the initial and final phases.

$$\Delta S = S_{final} - S_{initial} = R\left(x\ln(x) + (1-x)\ln(1-x)\right) < 0 \qquad (3)$$

Thus, the reaction will result in a decrease of entropy (the precipitation of A results in a transition from disorder to order). On the other hand, if an element B is soluble in $CoSi_2$ and insoluble in CoSi, the reaction is given by

$$(1-x)CoSi + xB + (1+x)Si \longrightarrow Co_{1-x}B_xSi_2 \qquad (4)$$

and in this case the difference in mixing entropy $\Delta S > 0$. For the intermediate case, in which some element C is partly soluble in CoSi and CoSi$_2$, ΔS $\neq 0$ if the solubility of C is different in CoSi and CoSi$_2$. The difference in mixing entropy ΔS will influence the change in free energy $\Delta G = \Delta H - T\Delta S < 0$, and according to eqn. (1) this will influence the activation energy for nucleation ΔG^*. Since Fe and Ge are more soluble in CoSi than in CoSi$_2$, one can expect that adding these elements to the silicidation reaction will heighten the nucleation barrier and thus increase the temperature of nucleation. In case of Ni, which is more soluble in CoSi$_2$ than in CoSi, a decrease of the nucleation temperature can be expected.

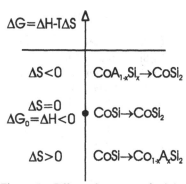

Figure 5 : Effect of entropy of mixing on $\Delta G = \Delta H - T\Delta S$.

THE ADDITION OF OTHER ELEMENTS

In binary alloys, an empirical rule by Hume-Rothery states that two elements are soluble if they have the same crystallographic structure and if their lattice parameter differs by less than 15%. One may expect that for silicides, tendencies for solubility may also be predicted based on arguments concerning crystallographic structure. In table I, an overview is given of the crystallographic structure of the different cobalt silicide phases, together with some other silicides that have a compatible crystallographic structure [6]. Although symmetry arguments can be used to predict whether the formation of a solid solution is possible, the heat of formation of the different silicide phases is important to predict effective solubility. This is illustrated by our measurements for Fe and Ge, where it was observed that Fe and Ge are expelled from the monosilicide, because H^f (CoSi) $<<$ H^f (CoGe) and H^f (CoSi) $<$ H^f (FeSi).

In this study, we have chosen Fe, Ge and Ni as alloying elements to illustrate the principle of influencing CoSi$_2$ nucleation by controlling the entropy of mixing. In the case of Ge and Ni, the results obtained in this work may have relevance for applications : Ge has potential applications for CoSi$_2$ formation on SiGe substrates, while the mixed Co$_{1-x}$Ni$_x$Si$_2$ may offer a possible alternative to CoSi$_2$ and NiSi for VLSI applications. Currently, we are studying the Co$_{1-x}$Ni$_x$Si$_2$ system in more detail. So far, we have found some interesting properties : the resistivity and thermal stability are comparable with that of CoSi$_2$, while the temperature of formation (400-500°C) is comparable to that of NiSi. Based on these properties, it seems that Co$_{1-x}$Ni$_x$Si$_2$ inherits the positive characteristics of both CoSi$_2$ and NiSi, while the negative aspects of both "parent" materials are avoided. Because of this, Co$_{1-x}$Ni$_x$Si$_2$ is a promising candidate for future SALICIDE processes. Currently, experiments are undertaken to implement this material for sub 0.25μm CMOS fabrication [7].

	Symmetry	Prototype	a [Å]	b [Å]	c [Å]	-ΔH [kcal/mole]
Co_2Si	Orthorhombic	Co_2Si	4.918	3.737	7.109	27.6
Ni_2Si	Orthorhombic	Co_2Si	7.060	4.990	3.720	31.5
Ru_2Si	Orthorhombic	Co_2Si	5.279	4.005	7.418	?
Rh_2Si	Orthorhombic	Co_2Si	5.408	3.930	7.383	18.4
Ir_2Si	Orthorhombic	Co_2Si	7.615	5.284	3.980	?
CoSi	Cubic	FeSi	4.447			24
CoGe	Cubic	FeSi	4.637			13.7
FeSi	Cubic	FeSi	4.490			19.2
CrSi	Cubic	FeSi	4.620			19
MnSi	Cubic	FeSi	4.548			26.6
RuSi	Cubic	FeSi	4.703			16
RhSi	Cubic	FeSi	4.685			16.2
ReSi	Cubic	FeSi	4.775			10.2
OsSi	Cubic	FeSi	4.729			15.6
$CoSi_2$	Cubic	CaF_2	5.364			24.6
$NiSi_2$	Cubic	CaF_2	5.406			20.7

Table I : Overview of the crystallographic structure and heat of formation of the different cobalt silicide phases, together with some other silicides with compatible crystal structures.

ACKNOWLEDGEMENTS

The authors would like to thank L. Van Meirhaeghe for technical support and U. Demeter for XPS measurements. C. Detavernier thanks the 'Fonds voor Wetenschappelijk Onderzoek – Vlaanderen' (FWO) for a scholarship. K. Maex is a research director for the FWO. A. Steegen and R.A. Donaton are acknowledged for assistance during the Nomarski measurements.

REFERENCES

1. F.M. d'Heurle, C.S. Petersson, Thin Sol. Films **128**, 283 (1985).
2. A. Appelbaum, R.V. Knoell, S.P. Murarka, J. Appl. Phys. **57**, 1880 (1985).
3. F.M. d'Heurle, J. Mater. Res. **3**, 167 (1988).
4. F.M. d'Heurle, D.D. Anfiteatro, V.R. Deline, T.G. Finstad, Thin Solid Films **128**, 107 (1985).
5. F. Wald, J. Michalik, J. Less-Common Metals **24**, 277 (1971).
6. K. Maex, M. Van Rossum, Properties of Metal Silicides, INSPEC, 1995, p.283.
7. K. Maex, MRS Proc. Spring Meeting 2000 (also in this volume).

Mat. Res. Soc. Symp. Vol. 611 © 2000 Materials Research Society

Effect of a thin Ta layer on the C49-C54 transition.

F. La Via[*], F. Mammoliti[#], M.G. Grimaldi[#], S. Quilici[°] and F. Meinardi[°].

[*] CNR-IMETEM, Stradale Primosole 50, Catania, Italy.
[#] INFM and Physics Department, Corso Italia 57, Catania, Italy.
[°] INFM and Material Science Department, Via Cozzi 53, Milano, Italy.

ABSTRACT

The effect of a thin Ta layer at the Ti/Si interface on the kinetic of the C49-C54 transition will be shown in detail. The transformation kinetic has been monitored by in situ sheet resistance measurements that, coupled to structural characterisation, allowed to evidence the presence of an intermediate phase before the C54 formation. The temperature of the C54 phase formation decreases with a Ta concentration of $4.5 \cdot 10^{15}$ cm^{-2} and μ-Raman images of partially transformed samples indicates that the density of C54 grains in presence of Ta is about one order of magnitude higher with respect to pure Ti/Si samples.

INTRODUCTION

TiSi$_2$ is widely used as a self-aligned silicide (salicide) for the metallization of gates, sources/drains and local interconnects in Ultra Large Scale Integration (ULSI) devices and circuits. In thin films, there are two crystalline phases with the TiSi$_2$ stechiometry: the high-resistivity (60-70 $\mu\Omega\cdot$cm) orthorhombic base-centred C49-TiSi$_2$ which usually forms at 550-700 °C and the low resistivity (15-20 $\mu\Omega\cdot$cm) orthorhombic face-centred C54-TiSi$_2$ phase, which usually forms at 700-850 °C [1]. The transformation from C49 to C54 phase becomes more difficult and requires annealing at high temperature in narrow (submicron) Si lines because of the low density [2] of nucleation sites.
In recent papers it has been shown that a significant reduction of the formation temperature of the C54 phase occurs when Mo is added to the reacting film by ion implantation [3] and/or deposition [4] of a thin layer at the Si/Ti interface. A similar effect has been observed even with Ta [5] or Nb [6] impurities and it has been attributed to the avoidance of the intermediate C49 phase during annealing since this phase was not detected by X-ray diffraction. However, in spite of the large

effort devoted to determine the effect of these metals on the C54 formation, a detailed kinetic study of the transformation is missing.

In this paper the effect of a thin Ta layer at the Ti/Si interface on the kinetic of the C49-C54 transition will be shown in detail. The transformation kinetic has been monitored by in situ sheet resistance measurements in samples with different Ta content. Structural characterization by X-ray diffraction and μ-Raman spectroscopy allowed detection of an intermediate phase before the C54 formation.

EXPERIMENTAL DETAILS

Samples were prepared in a UHV chamber by e-gun deposition of a 150 nm amorphous Si layer on thermally oxidised 5 inches silicon wafers. After this deposition, without breaking the vacuum, thin

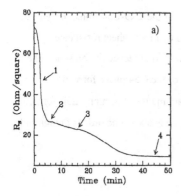

layers of Ta and Ti were deposited sequentially with the same technique. The Ta and Ti layers were 1.9×10^{15} at/cm^2 and 20 nm thick, respectively, as measured by Rutherford Backscattering Spectroscopy. In one case Ta was not deposited while the thickness of the Ti layer was maintained at 20 nm. Sheet resistance measurements were performed during the annealing in the temperature range between 500-650 °C at a pressure of $1\cdot10^{-7}$ Torr. Structural characterisation was performed on some selected samples by Transmission Electron Microscopy (TEM), Atomic Force Microscopy (AFM), X-Ray Diffraction (XRD), Rutherford Backscattering Spectroscopy (RBS) and μ-Raman. These samples were obtained stopping the annealing when the sheet resistance reached some characteristic values.

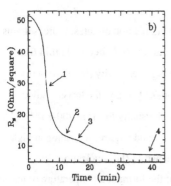

Fig. 1 - In situ sheet resistance measurement of a) Si/Ti sample annealed at 590 °C, b) Si/Ta/Ti sample annealed at 520 °C. The numbers reported in the figure show where the partially reacted samples were stopped.

RESULTS AND DISCUSSIONS

The sheet resistance versus time curve recorded during annealing at 590 °C of the sample without Ta is shown in Fig. 1a. It must be pointed out that the temperature of the sample was raised slowly in such a way that the sample temperature was stable at the origin of the time axis. The initial sheet resistance at t=0 s is due to the unreacted Ti film deposited on top of the amorphous Si layer. In about a minute the sheet resistance start decreasing with a high rate until it reaches a value of about 25 Ω/square after 5 minutes annealing. During the successive annealing the sheet resistance decrease is less steep and an almost flat region appears in the sheet resistance signal. At longer annealing times the sheet resistance decreasing rate increases again until a steady state value at about 10 Ω/square is reached. The arrows in the figure indicate the points at which the annealing was stopped for ex-situ characterisation. The in situ sheet resistance curve of the sample having 4.5×10^{15} cm^{-2} Ta at the Ti/Si interface is shown in fig.1b. The transition has been performed at 520 °C. Even in this case the resistance versus time curve shows an initial fast decrease followed by a more flat region and, at longer times, the final decrease to the steady state value of ~7 Ω/square. The difference of the sheet resistance absolute values measured in the two samples can be ascribed to minor difference in the silicide thickness and to the presence of Ta. However in presence of Ta the flat region is badly defined since it last for a relatively short time and the resistivity decreasing rate is almost constant. One more difference is related the absolute value of the sheet resistance of the plateau. In fact, the ratio of the sheet resistance of the sample with Ta to that without Ta is close to 0.75 before and at the end of the reaction whilst it is ~0.55 at the plateau. This could be an indication that a different intermediate phase is formed before the C54 phase.

The X-ray diffraction of partially transformed samples in which the annealing was stopped at the end of the flat region (point 3) are reported in Fig. 2 (top) for the pure Ti/Si case (continuous line) and for the sample with the Ta layer at the Ti-Si interface (dashed line). The spectra were taken in the Seeman-Bohlin (Grazing Angle) configuration. The X-Ray beam was incident at 1° on the sample surface and then planes that are parallel to the sample surface are not visible. Three well defined peaks at 28.5°, 41.0°, 47.5°, which can be attributed to Si (111), C49 (311), Si (220) , respectively, are visible in the diffraction pattern of the sample without Ta. Therefore the plateau occurring in the resistance versus time curve is due to the formation

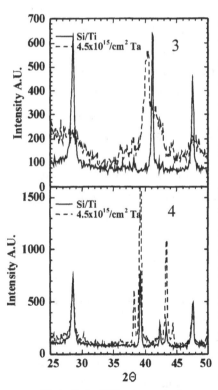

Fig. 2 - X_Ray Diffraction results for samples without and with Ta at the Si/Ti interface. Both the samples stopped in position 3 (upper figure), and the final samples (lower figure) are reported.

of the C49 phase. Crystallization of the amorphous Si occurs following the heat treatment. From these measurements it is possible to observe that in the first sheet resistance drop the C49 phase forms in the sample without Ta. The C49 amount continues to increase until the formation of the sheet resistance plateau. In that region a small decrease of the sheet resistance is observed and finally the C54 phase start to form after the second drop of the sheet resistance. A similar behaviour is observed also in the samples with a thin Ta layer at the Ti/Si interface. In this sample after the first sheet resistance drop the $TaSi_2$ (110) peak appears (not shown). When the sheet resistance signal reach the plateau, a wide peak appears clearly between 40 and 43 degrees. This peak is due to the overlay of two peaks: one is at 40.1°, where there is the peak $(2\bar{1}1)$ of the $TaSi_2$, and the other one is a very broad peak with the maximum close to 42°. A precise position of this last peak is difficult to be determined with only this measurement and more extended measurements have to be performed in order to determine the formed phase.

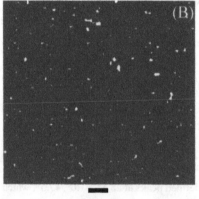

5 µm

Two different hypotheses can be done. Following the first one, reported also by S. Ohmi et al. [7], the C49 phase forms also in the sample with a Ta layer at the Si/Ti interface. In the second one, reported by A. Mouroux et al. [8], a ternary phase $(Ti,Ta)Si_2$ with a C40 structure forms before the C54 producing a faster transition to the final phase because of the C54 epitaxial growth on the ternary phase. From our measurements both these hypothesis can be excluded. In fact, the formation of the C49 does not occur as clearly shown by the in-situ sheet resistance measurements (Fig. 1) and by the XRD analyses of the partially reacted samples reported in figure 2 (upper). Furthermore, RBS spectra (not reported in this paper) show that the only reacted phases are $TiSi_2$ and $TaSi_2$. No ternary phases are formed. Then, during the reaction between the Ta/Ti bilayer and silicon a new intermediate phase, with an average stechiometry of $TiSi_2$, has been formed. More work should be done to understand the structure of this new phase.

Fig. 3 – µ-Raman images of the samples without (a) and with (b) Ta at the Ti/Si interface. The white areas are the C54 grains.

Continuing the reaction the C54 phase forms and the C54 peaks of the (311), (040), (022) and (331) planes appears for the sample without Ta. Instead, when Ta is introduced at the Si/Ti interface, the (311) and the (022) peaks grow considerably and the (131) peak appears. Also the (112) peak of the $TaSi_2$ has been detected in the final sample close to 44°.

The μ-Raman images reported in figure 3 show the samples without (Fig. 3a) and with Ta (Fig. 3b) after an anneal to obtain a C54 fraction between 4 and 1%. From the comparison it is clear that the introduction of a thin Ta layer at the Ti/Si interface produce an increase of the C54 nucleation centres. In fact, while in the reaction of a titanium layer on silicon a C54 nucleation sites density of about 0.01 $/\mu m^2$ is measured, in agreement with previous measurements [2], when a thin Ta layer is deposited at the Ti/Si interface the C54 nucleation sites density increases to 0.12 $/\mu m^2$. This increase is extremely important for the application of $TiSi_2$ in sub-micron devices because the use of the tantalum layer can decrease the minimum converted area of an order of magnitude.

CONCLUSIONS

In conclusion, the effect of a thin Ta layer at the Si/Ti interface on the C54 formation has been studied in detail. The transition occurs at lower temperature (about 70 °C) when there is a Ta layer at the Si/Ti interface. A new intermediate phase with $TiSi_2$ stechiometry has been detected during the reaction of the Si/Ta/Ti system. This phase has a lower resistivity (44 μΩ·cm) than that one of the C49 (84 μΩ·cm). μ-Raman analysis gives indications that the faster reaction of the samples with Ta at the Si/Ti interface is related to the increase of the C54 nucleation sites of an order of magnitude.

REFERENCES

[1] R. Beyers and R. Sinclair, J. Appl. Phys. 57, 5240 (1985).
[2] S. Privitera, F. La Via, M.G. Grimaldi and E. Rimini, Appl. Phys. Lett. 73(26), 3863 (1998).
[3] R.W. Mann, G.L. Miles, T.A. Knotts, D.W. Rakowski, L.A. Clevenger, J.M.E. Harper, F.M. D'Heurle, C. Cabral Jr. Appl. Phys. Lett. 67, 3729 (1995)
[4] A. Moroux, S.L. Zhang, W. Kaplan, S. Nygren, M. Östling and C.S. Peterson, Appl. Phys. Lett. 69(7), 975 (1996).
[5] A. Mouroux, S.L. Zhang and C.S. Petersson, Physical Review B 56(16), 10614 (1997).
[6] C. Cabral Jr., L.A. Clevenger, J.M.E. Harper, F.M. d'Heurle, R.A. Roy, C. Lavoie,, G.L. Miles, R.W. Mann and J.S. Nakos, Appl. Phys. Lett. 71, 3531 (1997).
[7] S. Ohmi and R.T. Tung, J. Of Appl. Phys., 86(7),3655 (1999).
[8] A. Mouroux, T. Epicier, S.L. Zhang and P. Pinard, Phys. Rev. B 60(12), 9165 (1999).

Mat. Res. Soc. Symp. Vol. 611 © 2000 Materials Research Society
SILICIDATION OF TITANIUM-RICH TITANIUM BORIDE DEPOSITED BY CO-SPUTTERING ON Si (100)

G. SADE AND J. PELLEG
Department of Materials Engineering, Ben-Gurion University of the Negev, Beer-Sheva, 84105, Israel

ABSTRACT

Titanium boride is known as a good diffusion barrier, in particular against copper, however outdiffusion of boron might deteriorate the semiconductor device. A $TiSi_2$ sublayer prevents effectively boron penetration into the Si substrate. In this study the intention was to form a $TiB_2/TiSi_2$ bilayer film by silicidation of a titanium-rich titanium boride deposited by magnetron co-sputtering from elemental targets. The $TiSi_2$ formation as well as the redistribution of titanium in the boride layer has been investigated by X-ray diffraction (XRD), Auger depth profiling and cross-sectional transmission electron microscopy (XTEM). Contact structure with Cu metallization was prepared to characterize this structure electrically.

The Ti-rich titanium boride film was completely amorphous by XRD up to 700 °C. Crystallization of Ti-rich silicides (Ti_3Si, Ti_5Si_3) have started at 750 °C, but already at 800 °C crystallization of C54 $TiSi_2$ was completed. TiB_2 begins to crystallize at 800 °C. Sheet resistance measurements confirmed these results. The sheet resistance of the as-deposited film was about 16 $\Omega/\square$ and no significant change was detected up to 700 °C. Then, a remarkable drop in the sheet resistance to ~1 $\Omega/\square$ was obtained after 800 °C, and this value was actually unchanged up to 925 °C. Cross-sectional TEM revealed the formation of the C54 $TiSi_2$ layer between TiB_2 and Si and additionally, a second C54 $TiSi_2$ layer was observed within the boride film. Current-voltage measurements of the prepared contact structure showed that it was a Schottky diode with very high leakage current.

INTRODUCTION

Recently, demands have been significantly increased to develop new interconnection materials in ultra large-scale integration (ULSI) devices. As a result, Cu is being considered as a potential substitute for Al because of its lower resistivity (1.67 $\mu\Omega$ cm) and higher resistance to electromigration [1, 2]. However, Cu diffuses easily into the Si substrate and SiO_2 layers at elevated temperatures during device fabrication and it reacts with Si to form Cu_3Si compounds at very low temperatures (200 °C) [3, 4]. Moreover, Cu acts as deep-level contaminant in Si that has an adverse effect on device performance. Thus, in order to apply the Cu interconnections in manufacturing devices, a diffusion barrier to prevent the Cu penetration into the Si substrate must be developed.

Various materials have been investigated as a diffusion barrier between Cu and Si such as transition metals [5, 6], intermetallic compounds [7], carbides [8], nitrides [9, 10] and ternary amorphous diffusion barriers [11, 12]. Wang [13] summarized in detail some of these barriers against Cu and their properties. Because of the excellent physical properties of TiB_2 [14, 15] it has been considered as a candidate for barrier application. Although this possibility was suggested as far as 1969 [16], not so many works have been reported since then on using TiB_2 for this application. Nicolet [17] has indeed suggested TiB_2 as a possible diffusion barrier in silicon technology. Blom *et al.* [18] and Shappirio *et al.* [19] examined the use of TiB_2 thin film as a diffusion barrier against Al, and Choi *et al.* [20] examined its applicability against Cu diffusion. No interaction took place between sputter deposited Cu and low-pressure chemical deposited amorphous TiB_2 film in vacuum anneals up to 750 °C for 30 min. However, Choi et al. [21] observed very strong outdiffusion of boron into Si from TiB_2, and they concluded that direct use of TiB_2 as an electrical contact to Si is limited. The remedy of B penetration into Si is to prepare a $TiB_2/TiSi_2$ bilayer [22, 23]. The $TiSi_2$ sublayer serves as an additional diffusion barrier against boron, since the boron diffusivity in $TiSi_2$ is very low up to 900 °C [24]. In our previous works the $TiSi_2$ sublayers were formed:1) by co-sputtering from elemental Ti and Si targets, followed by TiB_2 deposition [22, 23], and 2) by silicidation of TiB_2/Ti bilayer films [25].

The purpose of the present study is to evaluate the possibility of $TiB_2/TiSi_2$ bilayer formation from Ti-rich titanium boride deposited by co-sputtering. We expect that due to the small solubility of Ti in TiB_2, phase separation will occur resulting in Ti segregation at the TiB_2/Si interface. Ti will react with Si forming a $TiSi_2$ interlayer, which will retard B penetration into Si.

EXPERIMENT

N-type Si (100) wafers, 3 inch in diameter, were used as substrates. The substrates, before loading into the deposition chamber were dipped into a dilute 5% HF solution for 1 min, followed by rinsing in deionized water, acetone and alcohol, and finally they were r.f. plasma etched to remove native oxide. The base pressure in the chamber was <30 μPa. The high purity Ar (99.999%) introduced into the chamber was maintained at a pressure of about 0.5 Pa. Ti-rich titanium boride was deposited onto a rotating Si (100) substrate by co-sputtering from elemental Ti (99.999%) and boron (99.9) targets. Boron was sputtered with 320 W r.f. and Ti with 120 W d.c. power for 45 min providing a B/Ti ratio of ~1. The thickness of the as-deposited film was 120 nm. Post deposition annealing in a vacuum furnace in the range of 400 - 925 °C was used to induce $TiB_2/TiSi_2$ bilayer formation. The electrical resistivities of the samples were measured by the four-point probe method. A sample annealed at 800 °C and having the lowest sheet resistance was chosen for Schottky diode preparation and for detailed structural characterization by Auger electron spectroscopy (AES) and by XTEM. The Schottky diodes were prepared according to the procedures outlined by Choi et al. [21]. An active area of 250 x 250 μm^2 was defined by photolithography and wet etching. A second masking step was required to define the Ti-B film, and finally, after vacuum annealing a third masking step was used for the Cu layer. The effect of heating in the range of 400 - 700 °C on the properties of the contact was analyzed by I-V measurements. The structural studies were performed by XRD and XTEM, and the redistribution of the elements was analyzed by AES.

RESULTS AND DISCUSSIONS

Fig. 1 shows XRD spectra of the Ti-rich titanium boride film after vacuum annealing at the temperatures indicated. It is seen that the film is completely amorphous up to 700 °C, and only the forbidden Si (200) peak from the substrate is detected. At 750 °C Ti-rich silicides (Ti_3Si and Ti_5Si_3) and a hump of amorphous TiB_2 are seen. At 800 °C crystallization of C54 $TiSi_2$ was actually completed, but small quantities of Ti_5Si_3 also remained. Crystallization of TiB_2 is also initiated at this temperature. Annealing at 850 and 950 °C induces grain growth of TiB_2 and the remaining Ti_5Si_3 is transformed to C54 $TiSi_2$. The data of sheet resistance measurements (Fig. 2), confirm the results of XRD. It is seen that the sheet resistance of the as deposited film decreases gradually from 16 to 12 $\Omega/\square$ on annealing up to 700 °C. The drop in sheet resistance beyond 700 °C is a consequence of $TiSi_2$ and TiB_2 formation. It is worthwhile to note that after 800 °C the sheet resistance is nearly constant.

Figs.3 and 4 show I-V curves of the contact with Cu metallization after heat treatment in vacuum at the temperatures indicated. It is seen from the graphs that the contact behaves like a

Schottky diode, but having a very high leakage current even after low temperature annealing. Some improvement was observed at the high temperature annealing.

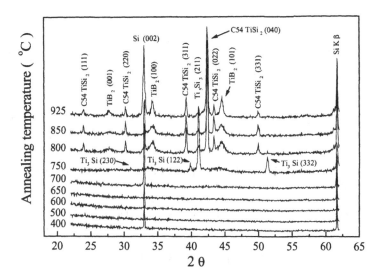

Fig. 1. XRD spectra of the as deposited Ti-rich titanium boride films on Si (100) substrate after annealing in vacuum for 30 min.

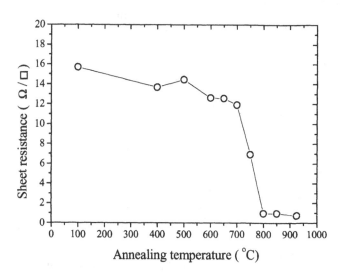

Fig. 2. Sheet resistance of the bilayer film as a function of annealing temperature.

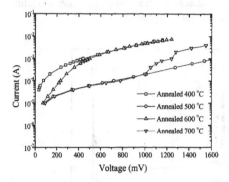

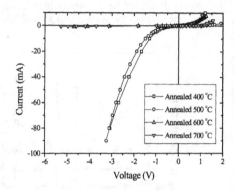

Fig. 3. Forward *I-V* characteristics of the contacts with Cu metallization after annealing.

Fig. 4. Reverse *I-V* characteristics of the contacts with Cu metallization after annealing.

Schottky barrier height, $\phi_{B,}$, of the heat-treated diodes were determined [26] using equation (1). These values as well as the leakage current at a reverse voltage of –5 V are listed in Table I.

$$J_s = A^{**}T^2 \exp(-q\phi_B / kT)[\exp(qV / nkT) - 1], \tag{1}$$

Here A^{**} is the effective Richardson constant (120 A/cm^2 K$^{\circ}$ for n-Si), T is the temperature, q is the charge of the electron, k is Boltzman's constant, and V is the applied voltage.

Table I. Calculated values of ϕ_B and the reverse leakage current at –5V.

Annealing temperature ($^{\circ}$C)	Schottky barrier height (ϕ_B,V)	Leakage current (mA)
400	0.63	200
500	0.70	300
600	0.68	1
700	0.72	0.8

The Schottky barrier heights are close to the value of TiSi$_2$ on n-Si [27], which was given as 0.6 V.

Auger depth profiling performed to study the redistribution of the elements in the film after heat treatment at 800 $^{\circ}$C (Fig. 5) shows that the TiSi$_2$ layer was formed between the titanium boride and the Si substrate, however a silicon layer was detected within the boride layer.

Cross-sectional images of this sample are shown in Fig. 6. Fig. 6a is a bright-field image of the sample.

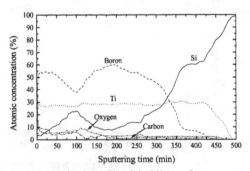

Fig. 5. Auger depth profile of the bilayer after vacuum annealing at 800 $^{\circ}$C for 30 min.

Electron diffraction from the TiB$_2$ area is shown in Fig.6d. At least three layers are seen in the structure: 1) at the Si-film interface - C54 TiSi$_2$, which also contains a small amount of Ti$_5$Si$_3$ in the vicinity of the TiB$_2$ layer, 2) intermediate region - amorphous TiB$_2$ and 3) external region - mixture of TiB$_2$ with amorphous Si. Lattice image of the TiSi$_2$/Si interface (Fig.6b) shows that it is rather smooth and does not contain any other phase, except TiSi$_2$ and Si. In some areas there might be a direct contact between amorphous TiB$_2$ and Si as it can be seen in the left lower corner of Fig. 6a. The presence of Si in the external layer indicates that as a result of annealing not only silicidation occurred at the Si/film interface but also Si diffusion happened through the layer to form a mixture of amorphous Si with TiB$_2$ crystals.

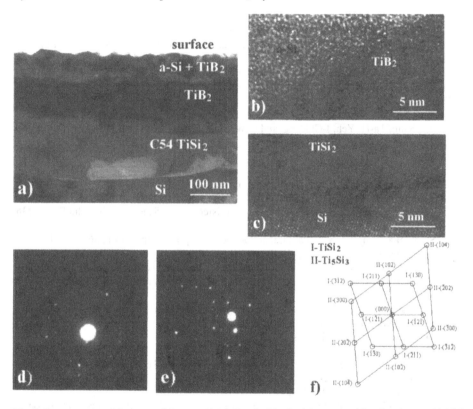

Fig.6. Cross-sectional images of the sample shown in Fig.5: a) low magnification image, b) high-resolution image of the surface layer, c) high-resolution image of the TiSi$_2$/Si interface, d) electron diffraction pattern from a selected area showing TiB$_2$, e) electron diffraction pattern from a selected area showing TiSi$_2$ and Ti$_5$Si$_3$, f) schematic diagram of the electron diffraction pattern of the TiSi$_2$ and the Ti$_5$Si$_3$.

CONCLUSION

TiB$_2$/TiSi$_2$ bilayer was prepared by silicidation of Ti-B film (B/Ti=1) deposited on Si substrate. Vacuum annealing up to 700 °C for 30 min does not change the amorphous state of the film. At 750 °C the crystallization of Ti-rich silicides was detected. Annealing at higher temperatures induces crystallization of C54 $TiSi_2$ and TiB_2 simultaneously. Contacts fabricated on this structure have a Schottky diode I-V characteristic, but with a very high leakage current. Auger depth profiling and XTEM revealed that the film after annealing at 800 °C for 30 min has transformed into a three-layer structure having a configuration of (a-Si+TiB_2)/(TiB_2)/(C54 $TiSi_2$+Ti_5Si_3). It may be noted that TiB_2/$TiSi_2$ prepared by other techniques [23, 25] provide better contact characteristics than the one obtained by the procedure described in this communication.

REFERENCES

1. C.W. Park and R.W. Vook, Appl. Phys. Lett. **59**, 175 (1991).
2. S.Q. Wang, I. Raaijmakers, B.J. Burrow, S.Suthar, S.Redkar, and K.-B. Kim, J.Appl.Phys., **70**, 5176 (1990).
3. S.Q. Hong, C.M. Comrie, S.W.Russell, and J.W. Mayer, J.Appl. Phys., **70**, 3655 (1992).
4. A. Cros, M.O.Aboelfotoh, and K.N. Tu, J.Appl. Phys. **67**, 3328 (1990).
5. S. Luby, E. Majkova, M. Jergel, M. Brunel, G. Leggieri, A. Luches, G. Majni, and Mengucci, Thin Solid Films, **277**, 138 (1996).
6. K.M. Chang, Ta-H. Yeh, I-C. Deng, and C.W. Shih, J. Appl. Phys., **82** 1469 (1997).
7. C.F. Hoener, E. Pylant, E.G. Boden, and S.-Q. Wang, J.Vac.Sci.Technol., **B 12**, 1394 (1994).
8. J. Imahori, T. Oku, M. Murakami, Thin Solid Films, **301** 142 (1997).
9. S.Dew, T. Smy, and M. Brett, Jpn. J. Appl. Phys., **33**, 1140 (1994).
10. M. Takeyama, A. Noya, T. Sase, and A. Ohta, J. Vac. Sci. Technol., **B 14(2)**, 674 (1996)
11. E. Kolawa, X. Sun, J.S. Reid, J.S. Chen , M.-A. Nicolet, and R. Ruiz, Thin Solid Films, **236**, 301 (1993).
12. Y. Shimooka, T. Iijima, S. Nakamura, and K. Suguro, Jpn. J. Appl. Phys., **36** 1589 (1997).
13. Shi-Qing Wang, MRS Bulletin, **19**(8), 30 (1994).
14. H. Holleck, Material selection for hard coatings, J. Vac. Sci. Technol., **A4**(6) 2661 (1986).
15. G. V. Samsonov and L. M. Vinitskii, Handbook of refractory compounds (Plenum Press, New York, 1979).
16. C.W. Nelson, 1969 Hybrid Microelectronics Symposium (International Society for Hybrid Microelectronics, Hicks Printing Co., Dallas, TX 1969).
17. M.-A. Nicolet, Thin Solid Films, **52**, 415 (1978).
18. H.-O. Blom, T. Larsson, S. Berg, and M. Ostling, J. Vac. Sci. Technol., **A7**,162 (1989).
19. J. R. Shappirio, J. J. Finnegan, and R. A. Lux, J. Vac. Sci. Technol., **B4**, 1409 (1986).
20. C. S. Choi, G. A. Ruggles, A. S. Shah, G. C. Xing, C. M. Osburn, and J. D. Hunn, J. Electrochem. Soc., **138**, 3062 (1991).
21. S. Choi, Q. Wang, C. M. Osburn, G. A. Ruggles, and A. S. Shah, IEEE Transactions on electron devices, **39**, 2341 (1992).
22. G. Sade, J. Pelleg, Mater. Res. Soc. Symp. Proc. **402**, edited by R. T. Tung, K. Maex, P. W. Pellegrini, and L. H. Allen (M.. R.. S. Symp. Proc. **402**, Pittsburgh, PA 1996) 131.
23. G. Sade, J. Pelleg and V. Ezersky, Microelectronic Engineering, **33**, 317 (1996).
24. P.Gas, V.Deline, F.M. d'Heurle, A.Michel, and G. Scilla, J. Appl. Phys., **60**, 1634 (1986).
25. G. Sade, J. Pelleg, Microelectronic Engineering, **37/38**, 535 (1997).
26. S. M. Sze, Physics of Semiconductor Devices, 2nd ed. (Wiley, New York, 1981) p. 245-311.
27. S. P. Murarka, Silicides for VLSI Application, (Academic Press, New York, 1983)p.15

Mat. Res. Soc. Symp. Vol. 611 © 2000 Materials Research Society

Thin Ni-Silicides For Low Resistance Contacts And Growth Of Thin Crystalline Si Layers

Elena A. Guliants and Wayne A. Anderson
Department of Electrical Engineering, State University of New York at Buffalo,
Buffalo, NY 14260

ABSTRACT

A new technological method of producing the Ni silicide with metal-like conductivity by deposition of a thin Si film over an ultrathin Ni prelayer at low temperature has been developed. The interaction of a metallic Ni with the Si atoms provided by the deposition source leads to the formation of the Ni-rich silicide phases immediately after the onset of Si deposition. Continued Si deposition results in the transformation of the Ni-rich silicide phases into the more Si-rich ones which implies that the phase composition is controlled by the Ni-to-Si concentration ratio rather than temperature. After Ni is completely consumed, the Si grains grow epitaxially on the disilicide crystals. The silicide layer has been studied in detail with respect to both the dynamics of the silicide growth and the electrical properties. The Ni silicide resistivity was found to be 2x 10^{-4} Ω-cm. The technique has advantages in two respects: it provides a high crystallinity Si film and allows fabrication of an ohmic contact directly on the substrate thus leaving the front surface of the film available for the formation of the active device junction.

INTRODUCTION

Metal silicides as ohmic contacts in Si-based technology introduced first in late 60s [1] continue to be a very topical subject for research [2,3]. $NiSi_2$ is among the most extensively studied silicides because of its perfect thermal stability and a small lattice mismatch with Si (0.4%) which makes Ni disilicide the best suited for the epitaxial growth on a Si substrate [4]. Nowadays, much effort is focused on the self-aligned silicidation process (SALICIDATION) where $NiSi_2$ is prepared by Ni deposition on a single crystal Si wafer followed by thermal annealing [5,6]. In such works, the Ni_XSi_Y phase formation and transformation was studied by varying the annealing temperature and, hence, was described in terms of thermodynamic processes as the well-known Ni_2Si-$NiSi$-$NiSi_2$ sequence [7]. The $NiSi_2$ was shown to grow epitaxially on c-Si.

This particular study is aimed at exploring the reverse process, namely silicon heteroepitaxy on $NiSi_2$, where the formation of Ni disilicide takes place during Si deposition on a Ni film. Recently, several reports were made on the annealing of the Ni-implanted a-Si, and the enhanced Si crystallization was observed in the vicinity of the formed $NiSi_2$ precipitates [8,9]. The metal-induced Si crystallization was observed also in the case of annealing of the a-Si covered with a thin Ni film [10,11]. Therefore, the technique introduced here may result in both enhanced Si crystal growth and self-organized and self-aligned $NiSi_2$, which serves as a satisfactory back ohmic contact for various planar microelectronic devices such as, for example, diodes and solar cells.

EXPERIMENT

A 200nm thick silicon oxide was grown on a single crystal Si wafer by plasma enhanced chemical vapor deposition (PECVD). This oxide layer was aimed at providing a sufficient diffusion barrier for Ni and thus preventing the silicide formation at the Ni/c-Si interface. Ni films with thickness of 25-30nm were deposited by thermal evaporation of a 99.99% Ni wire on the above substrates at a base pressure in the low 10^{-6} Torr range. The thickness of the Ni film was controlled with a quartz crystal thickness monitor with an approximate error of 5%, and the deposition rate was about 80Å/s. The Si deposition was performed by direct current (d.c.) magnetron sputtering from a 99.99% Si target at a base pressure of 2×10^{-7} Torr. The deposition was carried out in a 5%H_2/Ar mixture at a pressure of 1mTorr. The substrate temperature was kept at 575-600°C and the magnetron power was 50W which provided the deposition rate of 0.5μm/hour. Standard organic cleaning was performed before each successive fabrication step. In all instances, the substrates were cleaned by sequential ultrasonic processing for 2 minutes in acetone and methanol, rinsed in deionized water and dried in flowing nitrogen gas.

The structure of the resulting silicide layer was examined using cross-sectional transmission electron microscopy (XTEM) and x-ray difraction (XRD). The Ni silicide phase formation and transitions were studied by Rutherford backscattering spectrometry (RBS). Because of the inaccessibility of the RBS analysis for in-situ measurements, the Si deposition was repeated with identical parameters and interrupted sequentially at different stages of the silicide growth. The Ni silicide resistivity measurement are described in context below.

RESULTS AND DISCUSSION

Figure 1 shows a low magnification XTEM image of a 500nm Si film deposited on a SiO_X/Si substrate coated with 25nm of Ni at 600°C. The dark area represents a silicide layer formed as a result of the reaction between Ni and sputtered Si atoms. Silicon is more electron-transparent and represents the lighter area on the image. This layer exhibits a columnar structure confirming the metal-induced Si grain growth. A 25nm thick Ni film is seen to provide the thickness of the Ni silicide layer in the 70-120nm range which, taking into account that the Ni atoms are slightly larger than the Si ones, implies the Si-rich phase. A higher magnification XTEM image of the Si – Ni silicide interface is shown in Fig. 2. Selected area diffraction (SAD) patterns taken in the silicide region helped to identify a cubic $NiSi_2$ (a =5.416 Å), as seen, for example, in Figure 2. These patterns are predominantly from the low index poles, which is typical for the $NiSi_2$. The disilicide layer possesses a somewhat preferential crystal orientation demonstrating mostly [111] patterns. Single-crystal SAD patterns, which originated from this region, indicate large grain size implying that the variation in the thickness of the silicide is due to the morphology of isolated $NiSi_2$ grains.

In the XRD scan of the identical sample, a 500nm Si/25nm Ni/SiO_X/c-Si, only three peaks are present. These peaks observed in the vicinity of 28.5°, 47.5°, and 56.2° correspond to Si(111), Si(220) and Si(311) diffracted intensities, respectively. On the other hand, theoretically, due to a miniscule lattice mismatch, the peak positions of the corresponding diffracted planes for $NiSi_2$ are very close to those of Si. Subsequently, each of the diffuse peaks can be deconvoluted to two peaks: Si (at smaller values of 2θ) and Ni disilicide (at greater

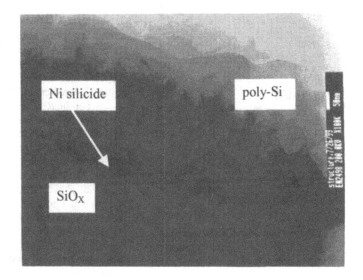

Figure 1. An XTEM image of a 500nm Si / 25 nm Ni / SiO$_X$ / Si sample
deposited at 600°C

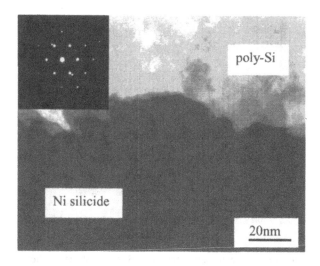

Figure 2. A high magnification XTEM image of the silicide layer with a [1 1 1] SAD pattern
taken at the center of the image, indicative of the NiSi$_2$ phase

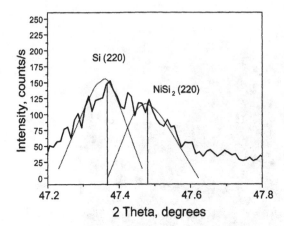

Si (220)

NiSi₂ (220)

values, as seen, for example, in Fig. 3, where the XRD scan is centered around Bragg angles of 23.5° - 24° ($2\theta = 47° - 48°$), in the vicinity of the (220) diffracted intensity. Thus, the XRD analysis confirms the epitaxial alignment of Si on NiSi$_2$.

For the RBS analysis, a 30nm thick Ni film was evaporated on the c-Si wafer, covered with a 300nm thick PECVD silicon oxide. The sample was then cut into three pieces. The first piece was studied to provide a backscattering spectrum of a pure, unreacted nickel. The second and the third pieces served as substrates for the Si films, deposited at standard deposition conditions and a substrate temperature of 600°C for 3 min and

Figure 3. The (220) portion of the XRD spectrum of a 25nm Ni/500nm Si/SiO$_X$/c-Si sample indicating the epitaxial alignment of Si on NiSi$_2$.

15 minutes, respectively. The random RBS spectra of all three samples were plotted in the same figure (Fig. 4) to allow a quantitative composition analysis of the formed phases.

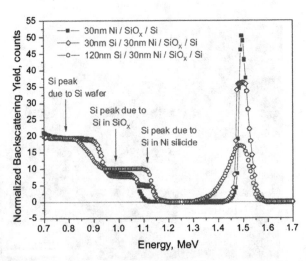

Figure 4. Rutherford backscattering spectra of a Si-on-Ni structure showing the sequential growth of Ni silicide phases. Solid symbols represent the Ni film on a SiO$_X$ / Si substrate prior to Si deposition.

The intensities of the Ni and Si peaks for the second sample are 36 counts and 4.5 counts, respectively. Taking into account the ratio of the squared atomic masses, the composition of the film is Ni : Si = $36 \times (14)^2 : 4.5 \times (28)^2 = 2 : 1$. Clearly, after 3.5 minutes of the Si deposition, all Ni reacted with sputtered Si atoms to form Ni_2Si. Analogously, the intensities of Ni and Si peaks in the third sample have the respective values of 16 counts and 10 counts, making the film composition Ni : Si = 1 : 2.5. This implies that the film is composed of the Ni disilicide, $NiSi_2$, plus a small amount of elemental Si, which emerges due to Si saturation in the silicide. Therefore, since these two samples were produced at the same temperature, it can be concluded that in the case of Si deposition on a Ni prelayer, the silicide formation is controlled by the Ni-to-Si concentration ratio rather than temperature.

The resistivity of Ni silicide was calculated from the slope of the current-voltage (I-V) characteristics taken for the structure shown in Fig. 5 which was prepared as follows. The standard sample was thoroughly cleaned in acetone, methanol and DI water. The silicon side was glued to a 10mmx2mm piece of an alumina (Al_2O_3) substrate with epoxy. The sample was then subjected to ultrasonic processing in 49% HF for several hours in order to etch away the silicon oxide. The c-Si substrate was removed, and the Ni silicide/polysilicon structure was obtained on an insulating substrate. The resistivity of polysilicon film was a priori measured to be in the $10^3\Omega$-cm range by taking vertical I-V characteristics on the initial sample. Thus, assuming that polysilicon exhibits insulating properties compared to the silicide and taking into account the silicide thickness obtained from XTEM data, the Ni silicide resistivity was found to be $1.5 \times 10^{-4}\Omega$-cm. This value is close to the previously reported values of $30\mu\Omega$-cm [5] and 130 $\mu\Omega$-cm [6] for $NiSi_2$.

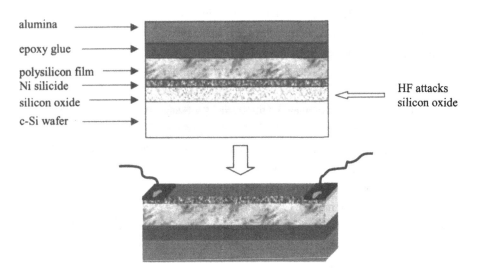

Figure 5. Schematic representation of the sample preparation for the silicide resistivity measurement.

CONCLUSIONS

Silicon was deposited on a thin Ni prelayer at 575-600°C and yielded a high crystallinity columnar structure. A 70-120nm thick Ni silicide layer was formed as an intermediate layer between the substrate and the Si film as a result of the chemical reaction between Ni and Si. This layer was identified as the $NiSi_2$ phase by both XTEM and XRD analysis. The Si film was found to be highly epitaxially aligned with $NiSi_2$. The silicide resistivity was found to be in the low $10^{-4}\Omega$-cm range making it suitable to use as a back ohmic contact in various Si-based devices. The technique has an advantage of the polycrystalline Si growth as well as the self-organized and self-aligned ohmic contact formation.

ACKNOWLEDGEMENT

The authors acknowledge the partial financial support by NASA and New York State Energy Research and Development Authority, and would like to thank Drs. J.Castracane and H.Efstathiadis at SUNY Albany for the RBS analysis.

REFERENCES

1. M.P.Lepselter and J.M.Andrews, in *"Ohmic Contact to Semiconductors"*, edited by B.Schwartz, *The Electrochemical Society, New York* (1969).

2. H.Föll, P.S.Ho, and K.N.Tu, J. Appl. Phys., **52**, 250 (1981).

3. R.Pretorius and J.W.Mayer, J. Appl. Phys., **81**, 2448 (1997).

4. Y.Z.Feng, Z.Q.Wu, J. Mat. Sci. Let. **15**, 2000 (1996).

5. F.Deng, R.A.Johnson, P.M.Asbeck, S.S.Lau, W.B.Dubbelday, T.Hsiao and J.Woo, J.Appl. Phys. **81**, 8047 (1997).

6. E.Maillard-Schaller, B.I.Boyanov, S.English, and R.J.Nemanich, J.Appl. Phys. **85**, 3614 (1999).

7. K.N.Tu and S.S.Lau, in *"Thin Films – Interdiffusion and Reactions"*, edited by J.M.Poate, K.N.Tu, and J.W.Mayer , *Wiley, New York* (1978).

8. G.Liu and S.J.Fonash, Appl. Phys. Lett. **55**, 660 (1989).

9. C.Hayzelden and J.L.Batstone, J.Appl. Phys. **73**, 8279 (1993).

10. S. Y.Yoon, K.H.Kim, C.O.Kim, J.Y.Oh and J.Jang, J.Appl. Phys. **82**, 5865 (1997).

11. Z.Jin, G.A.Bhat, M.Yeung, H.S.Kwok, and M.Wong, J.Appl. Phys. **84**, 194 (1998).

Mat. Res. Soc. Symp. Vol. 611 © 2000 Materials Research Society

REMOTE PLASMA NITRIDATION OF In-Situ STEAM GENERATED (ISSG) OXIDE

H.N. Al-Shareef, A. Karamcheti, T.Y. Luo, G.A. Brown,V.H.C. Watt, M.D. Jackson, and H.R. Huff, International Sematech, Inc., Austin, TX 78741

R. Jallepally, D. Noble, and N. Tam, and G. Miner, Applied Materials, Santa Clara, CA 95050

ABSTRACT

Electrical performance of in-situ steam generated (ISSG) oxide nitrided using remote plasma nitridation (RPN) has been evaluated. An equivalent oxide thickness (EOT) of 1.6 nm with gate leakage current around 5×10^{-3} A/cm^2 (at -1.5V) has been achieved. The leakage current of remote plasma nitrided ISSG oxide is lower than that of ISSG only, where more than one order of magnitude leakage current reduction (at the same EOT) has been achieved for some RPN conditions. Moreover, it is observed that the extent to which the RPN process conditions modify device parameters such as EOT, flatband voltage (V_{FB}), and time-to-breakdown (t_{bd}) increases with decreasing ISSG thickness. The thinner ISSG oxides appear to be more susceptible to plasma damage and accumulation of positively charged nitrogen atoms at the oxide/Si interface. Therefore, RPN processes that use lower temperature and shorter time are preferred for very thin oxides. The nitrogen content and profile in the samples evaluated using SIMS analysis, indicate that RPN offers higher nitrogen content and better nitrogen profile compared to conventional nitrogen incorporation methods such as NO annealing [1].

INTRODUCTION

Extensive efforts are currently underway to find alternative gate dielectrics for conventional silicon dioxide. The scaling of SiO$_2$ thickness will most likely continue down to the 150 nm technology node. However, the required SiO$_2$ thickness for the 130 nm node is projected to be 1.4-1.9 nm according to the International Technology Roadmap for Semiconductors (ITRS)[2]. At these thickness levels SiO$_2$ may not work properly due to excessive tunneling leakage currents and boron penetration from the p+ polysilicon electrodes. It therefore appears that alternative gate dielectrics will be needed for the 130 nm node, and several have already been investigated. These include a silicon oxide/silicon nitride stack [3], remote plasma nitrided (RPN) silicon oxide [4,5], and jet vapor deposited (JVD) nitride [6]. In this work, we have evaluated remote plasma-nitrided (RPN) in-situ steam generated (ISSG) oxide as a gate dielectric for the 130 nm technology node. The choice of the RPN process has been driven by the apparent RPN process ability to introduce large amounts of [N] into the SiO$_2$ film while carefully controlling the nitrogen depth profile [4]. However, care must be exercised as excessive amounts of [N] near the Si/SiO$_2$ interface can lead to mobility degradation.

EXPERIMENT

MOS capacitor devices with oxide isolation were fabricated on high quality 200 mm EPI p/p$^+$ (100) Si substrates. The wafers were cleaned in a 100:1 HF solution (DHF) prior to ISSG oxide growth. An AMAT Centura RTP system with in-situ steam generation capability was used to grow 1.6-2.1 nm thick ISSG oxides. The ISSG films were grown at 950°C in a 1%H_2 (99%O_2) ambient at 10 Torr. Nitridation of the ISSG oxides was carried out using remote high-density N_2 plasma using a N_2/He ratio of 1:4. The RPN process temperature was varied between 550-850°C while the RPN process time was either 150 or 240 seconds. Table 1 shows the details of the four RPN processes used. Undoped polysilicon gate electrodes were subsequently deposited at 720°C in an AMAT polysilicon Centura. Polysilicon doping and activation was performed prior to electrical testing using ion implantation and rapid thermal annealing.

The electrical measurements were performed using an HP4284 LCR meter for the C-V measurements and an HP 4156B parametric analyzer for the I-V measurements. The EOT values were extracted form the 100 kHz, parallel mode C-V characteristics using both the IBM TQM model [7] and the CVC model from North Carolina State University [8].

Table 1 Remote plasma nitridation (RPN) conditions used in this study

Process	Temperature (°C)	Time (sec)	Pressure (Torr)	%He (He/N_2)	Flow (slm)	Power
RPN1	750	240	1.8	80	3.0	3 kW
RPN2	750	150	1.8	80	3.0	3 kW
RPN3	550	240	1.8	80	3.0	3 kW
RPN4	850	150	1.8	80	3.0	3 kW

RESULTS AND DISCUSSION

The driving force for using the RPN process, as mentioned earlier, is the ability of the process to incorporate large amounts of nitrogen into the gate dielectric with good control over location of the nitrogen peak[4]. Fig.1 shows the nitrogen profile in a nominally 4.0 nm ISSG oxide nitrided using the RPN2 process. Notice that up to 15 atm% nitrogen is incorporated into this ISSG oxide. In addition, the nitrogen peak is located in the upper part of the gate dielectric and away from the ISSG oxide/Si interface. This nitrogen profile is nearly ideal as it offers the benefit of increased resistance to boron penetration without degrading carrier mobility.

On the other hand, the uniformity of the ISSG films following remote plasma nitridation can degrade depending on the RPN conditions used. Fig.2 shows the thickness range across 8" wafers for 18Å ISSG oxide subjected to four different RPN process conditions. Notice that the thickness range across the 8" wafer is mostly larger for the ISSG/RPN process than for pure ISSG oxide. In addition, the more severe RPN conditions (i.e., higher temperature and longer time as in RPN1 and RPN4) resulted in larger thickness range. This observation suggests that performing remote plasma nitridation at lower temperatures and for shorter times is preferred as it gives thickness uniformity comparable to pure ISSG. Additional work is needed to optimize the uniformity of remote plasma nitrided ISSG oxides before they can enter full production.

Fig.3 shows the gate leakage current versus equivalent oxide thickness (EOT) for ISSG and ISSG/RPN oxides. The figure clearly shows that subjecting the ISSG oxide to either RPN2 or RPN3 process reduces gate leakage current by up to 20X at the same EOT. It is important to note that the flatband voltage for all devices compared in Fig.3 have the same flatband voltage, making the comparison at −1.5V a valid one. Similar performance was observed in the case of RPN1 and RPN4 (not shown). Fig.3 also shows that the magnitude of leakage current reduction from the RPN process decreases with increasing equivalent oxide thickness. In fact, all the electrical parameters measured show a similar trend, where the RPN conditions have much less effect on ISSG oxide properties as the thickness of the starting ISSG oxide increases. For instance, Fig.4 shows the effect of four different RPN processes on the EOT uniformity for ISSG films with an initial thickness of 16 or 20 Å. Notice that the EOT of the 20Å oxide is the same for the ISSG and the ISSG/RPN oxides. We believe this result to be due to two competing effects. One effect comes from the fact that the RPN process increases the thickness of the

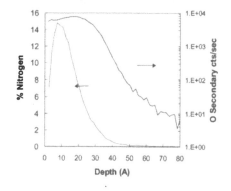

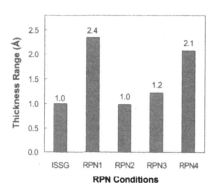

Figure 1 SIMS depth profile showing the location and atomic concentration of nitrogen in a 4.0 nm ISSG oxide.

Figure 2 Thickness range across 8" wafers with ISSG and ISSG/RPN dielectrics.

starting ISSG oxide . This statement is substantiated by the TEM picture shown in Fig.5 where an initially 1.6-1.7 nm thick ISSG oxide increases to around 2.6 nm after RPN1 (a rather severe RPN condition). The mechanism for this will be discussed later, but this should increase EOT of the stack. At the same time, nitrogen incorporation from the RPN process increases the effective dielectric constant of the stack, decreasing EOT in the process. These two effects roughly cancel and the resulting EOT of the 20Å ISSG oxide remains the same regardless of what RPN conditions are used.

In contrast, Fig.4 also shows that EOT of the ISSG/RPN dielectric does depend on the RPN process conditions for the 16Å ISSG oxide. If one starts with 16Å ISSG oxide, the observed EOT changes following RPN. We suggest that this behavior may be due to the nitridation of the Si substrate during the RPN process in the case of very thin ISSG oxide. This nitridation may lead to formation of an interfacial nitride layer underneath the 16Å ISSG oxide, which reduces the overall EOT despite increasing stack thickness. This is due to the higher dielectric constant of such a nitrided layer. This observation is in contrast to the 20Å ISSG oxide case, where the higher starting thickness of ISSG oxide precludes any substantial nitridation of the Si substrate. There is evidence for and against the idea of Si substrate nitridation. For example, the TEM picture shown in Fig.5 shows what appears to be a single dielectric layer rather than two. Had there bee a completely nitrided layer underneath the ISSG oxide, this would have been discernible in the TEM image. On the other hand, the flatband voltage (V_{FB}) values in Fig.6 show that remote plasma nitridation of 16Å ISSG oxide results in more negative flatband voltages than observed for pure ISSG. A more negative flatband voltage is usually indicative of positive fixed charge (e.g. positively charged N atoms) at the SiO_2/Si interface of these NMOS devices but may also result from nitridation of the Si substrate. Nitrides tend to produce more negative flatband voltages than pure oxide. We are currently performing XPS as a function of depth in the ISSG/RPN dielectric. This should provide bonding information at various depths and hopefully help explain the reason for thickness increase observed after remote plasma nitridation. The absence of this strong V_{FB} dependence on RPN conditions in the case of the 20Å ISSG oxide as shown in Fig.6 is indicative of less positive fixed charge or less nitride-like material at the interface. Thus it should be obvious from this flatband voltage and from the earlier leakage current and EOT data that the impact of RPN process on ISSG is strongly dependent on the starting ISSG oxide thickness.

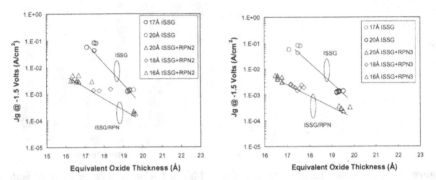

Figure 3 Gate leakage versus equivalent oxide thickness of ISSG and ISSG/RPN oxides.

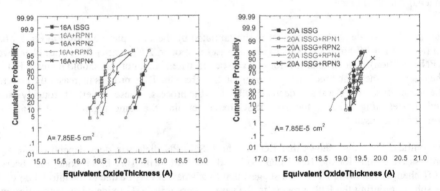

Figure 4 Cumulative probability plots showing EOT for various ISSG/RPN processes.

Preliminary reliability data have also been collected on these samples ad Figure 7 shows the effect of ISSG/RPN process conditions on the charge-to-breakdown values measured using the constant voltage method. It is observed for the two ISSG thicknesses (16 and 20Å) that the charge-to-breakdown decreases as the time and temperature of the RPN process increase. While we show data for only two stress conditions, we note that multiple stress voltage measurements have been performed on each sample and the data consistently show that the RPN conditions

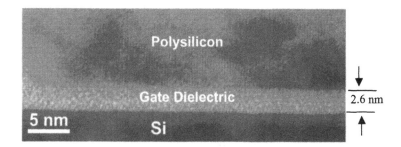

Figure 5 TEM cross section of an ISSG oxide after remote plasma nitridation. The oxide was initially 1.6-1.7 nm thick and increased to 2.6 nm after remote plasma nitridation.

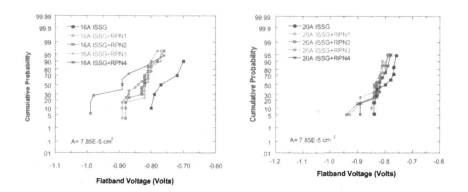

Figure 6 Flatband voltages of 16 and 20Å ISSG oxides before and after RPNb

used in this study degrade ISSG reliability. Similar to previous data, we also observe that the effect of the RPN process conditions on ISSG oxide reliability becomes more pronounced as ISSG thickness decreases. In this case, pure ISSG shows the best charge-to-breakdown while RPN3, which is the mildest of all RPN processes used (lowest temperature) has the next best charge-to-breakdown. As the RPN conditions become more severe, the charge-to-breakdown continues to degrade. The effect of RPN process on oxide reliability may be a result of introducing charging trapping centers in the oxide (e.g. positively-charged nitrogen atoms) which increase charge trapping in ISSG oxide, leading to reliability degradation.

Finally, in the data discussed so far the equivalent oxide thickness (EOT) was determined using the IBM TQM model [7]. However, another model, namely, the NC State University CVC model is also commonly used to extract EOT values from CV curve[8]. Fig.8 shows the correlation between CET (the capacitance value obtained from the CV curves without poly depletion and quantum mechanical corrections) and EOT obtained by both the IBM and NCSU models. The data points were measured on MOS capacitors with various gate dielectrics such as ISSG oxide, ISSG/RPN, oxynitride, oxide-nitride dielectrics. We observe a nearly constant difference of 1.8Å between the two models. This difference appears to be relatively independent of which dielectric process is used.

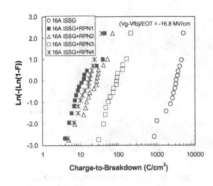 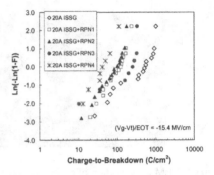

Figure 7 Time-to-breakdown of 16 and 18Å ISSG and ISSG/RPN gate dielectrics.

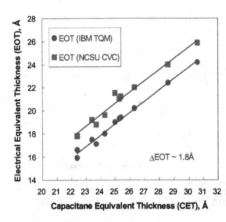

Figure 8 Correlation between equivalent oxide thickness (EOT) and capacitance equivalent thickness (CET) for various gate dielectric processes.

CONCLUSIONS

Remote plasma nitridation of ISSG oxide can substantially reduce ISSG oxide leakage. The choice of optimum RPN conditions depends strongly on the starting ISSG oxide thickness. For a given ISSG thickness, the choice of optimum RPN conditions is a tradeoff between two factors. On the one hand, an aggressive RPN process (high temperature and longer time) will reduce gate leakage and improve EOT. One the other hand, an aggressive RPN process may degrade oxide uniformity and introduce large amounts of positively charged nitrogen atoms at oxide/Si interface, causing flatband voltage shifts and degrading device mobility.

The authors would like to thank Gennadi Bersuker and Kenneth Torres at International Sematech and Khaled Ahmad at Conexant Systems for several useful comments.

REFERENCES
[1] D.A. Buchannan, IBM. J. Res. Develop. **43**,245(1999).
[2] ITRS, Semiconductor Industry Association, San Jose, CA 95129, 1999.
[3] S.C. Song et al., IEDM, p.373, 1998.
[4] S.V. Hattangady et al., IEDM Tech. Digest, p.495(1996).
[5] S.V. Hattangady, H. Niimi, and G. Lucovsky, Appl. Phys. Lett. **66**,3495(1995).
[6] T.P. Ma, Electrochemical Society Proceedings 99-10,, 57(1999).
[7] S.-H. Lo, D.A. Buchanan, and Y. Taur, IBM. J. Res. Develop. **43**,327(1999).
[8] Khaled Ahmed et. al., " Comparative Physical and Electrical Metrology of Ultra-Thin Oxides in the 6-1.5 nm Regime," Accepted for publication in the IEEE Transaction on Electron Devices.

Silicide Formation Mechanisms

Mat. Res. Soc. Symp. Vol. 611 © 2000 Materials Research Society

ON THE TEMPLATE MECHANISM OF ENHANCED C54-TiSi$_2$ FORMATION

L. KAPPIUS and R. T. TUNG*
Bell Labs, Lucent Technologies, 600 Mountain Ave. Murray Hill, N.J. 07974
* rtt@lucent.com

ABSTRACT

The enhanced formation of the C54-TiSi$_2$ phase by the addition of small amounts of refractory metal (Tm = Mo, Ta, Nb, ..) has often been ascribed to a template mechanism from the C40 Ti$_x$Rm$_{1-x}$Si$_2$ or the (Ti,Rm)$_5$Si$_3$ phase. Due to lattice matching conditions, the presence of either of these phases is thought to lower the interface energies with certain orientations of the C54-TiSi$_2$ grain and, thereby, possibly lower the nucleation barrier of the C54-TiSi$_2$ phase. These proposed template mechanisms are specifically tested in the present work through a study of the nucleation of TiSi$_2$ phase(s) in contact with a pre-existing C40 Ti$_{0.4}$Mo$_{0.6}$Si$_2$ or Ti$_5$Si$_3$ layer. No identifiable enhancement in the C54-TiSi$_2$ nucleation was observed which could be attributed to templates. Instead, the nucleation temperature of the C54-TiSi$_2$ phase appeared to be correlated with the grain size of the C49-TiSi$_2$ layer, independent of whether Rm was present. These results are suggestive that the primary mechanism for the enhanced formation of the C54 phase by refractory metals is a reduction in the grain size of the C49 TiSi$_2$ phase, likely due to altered kinetics.

INTRODUCTION

The difficulty with the polymorphic C49 -> C54 phase transformation in narrow TiSi$_2$ lines is the primary reason for the steady-paced phasing-out of Ti self-aligned silicide (salicide) technologies from processes for deep sub-micron ULSI devices. This difficulty lies usually in a low density of nucleation sites for the C54-TiSi$_2$ phase, which are believed to reside at multiple-grain nodes of the polycrystalline C49 TiSi$_2$ layers.[1] Recently, the incorporation of a small amount of refractory metal, e.g. Mo, Ta, Nb, etc, has been shown to lead to an enhancement in the nucleation of the C54 phase. Refractory metals introduced through ion implantation,[2] interposed layer,[3] and dilute Ti$_{1-x}$Rm$_x$ alloy [4] all led to a lowering of the C49 -> C54 transformation temperature, concurrently with the observation of a reduction in the grain size of the TiSi$_2$ phase(s) being formed. Based on powder x-ray diffraction (XRD) result, the enhanced formation of the C54 phase was once speculated to occur because the C49 TiSi$_2$ phase was by-passed in the silicide reaction. Later, transmission electron microscopy (TEM) analysis showed the presence of a fine-grain C49 layer prior to the formation of the C54 TiSi$_2$ phase and suggested that the absence of C49-related peaks in the XRD was due to the very small grain size of this phase, in the presence of refractory metal impurities.[5] It was also argued that since a major effect of refractory metal impurities was a decrease in the grain size of the C49 TiSi$_2$ layer, this effect alone could potentially account for the observed enhancement in the formation of the C54 TiSi$_2$ phase.[5] A more often discussed explanation of the enhanced nucleation of the C54 phase is based on a template mechanism via observed reaction by-products or transient products. A lattice matching condition between the (040) plane of C54-TiSi$_2$ and the (0001) plane of a hexagonal C40 Ti$_{1-x}$Rm$_x$Si$_2$ phase has been notices,[3] as has a matching condition between the C54-TiSi$_2$ (040) and the Ti$_5$Si$_3$ (300) planes.[6] Since a phase based on the Ti$_5$Si$_3$ has been observed as a precursor, and a C40 phase has been observed as a by-product during Ti-Si reactions involving Rm impurities, the lower energy of these epitaxial interfaces have been speculated to be responsible for the observed enhancement of the C54-TiSi$_2$ phase. Additionally, it has been proposed that Rm impurities tend to stabilize the C54 phase, thereby increasing the thermodynamic driving force for the transformation.[7] The present work searches for possible template action at interfaces between these proposed materials and TiSi$_2$. No clear evidence for template action could be found. Instead, results were consistent with the C54 transformation temperature being correlated with only the grain size of the C49 phase, independent of whether templates and/or Mo were present. They are also suggestive that the beneficial effect of Rm additives may be related to the reduced reaction speed.

EXPERIMENTAL

Si substrates with a 50-100nm thermal oxide layer were chemically cleaned, loaded in a UHV chamber, and degassed. Separate e-beam sources were used to deposit Ti, Mo, and Si, or to co-deposit layers of desired compositions at controllable substrate temperatures. Some samples were annealed and received further evaporation and annealing steps, all in UHV. Other samples were given only mild anneals in situ (to prevent oxidation) and were annealed ex situ in a rapid thermal annealing (RTA) tool. All the Ti silicide layers studied in this work have a final thickness of ~34nm. Samples were analyzed by Rutherford backscattering, TEM and 4-point probe measurement.

RESULTS AND DISCUSSION

Thin Ti$_{0.4}$Mo$_{0.6}$Si$_2$ template layers were grown on SiO$_2$ by co-deposition at room temperature and annealing at 580-650°C. TEM positively identified the formation of fine polycrystals of the C40 phase for layers with thickness μ 2nm. In one experiment, a layer of 15nm* thick amorphous TiSi$_{2.0}$ and a 10nm thick Si cap layer were deposited at room temperature on a 2nm Ti$_{0.4}$Mo$_{0.6}$Si$_2$ thick template layer and were given a brief anneal at 550°C in UHV. The sample was then cleaved into pieces and RTA'd at different temperatures. The sheet resistance of silicide layers processed this way is marked "Mo on bottom" in Fig. 1, where comparison is made with samples which were processed similarly but without the Ti$_{0.4}$Mo$_{0.6}$Si$_2$ template, marked "pure TiSi$_2$". Also shown in Fig. 1, marked "Mo on top", is the result for a 34nm thick, essentially C49-TiSi$_2$, layer, grown by the depositions of 10nm Si and 15nm* TiSi$_{2.0}$ on SiO$_2$ and a brief anneal at 550°C. A 2nm Ti$_{0.4}$Mo$_{0.6}$Si$_{2.0}$ layer was then deposited on top of the TiSi$_2$ layer and given a brief anneal at 550°C. From Fig. 1, the C54 transformation temperature for TiSi$_2$, which is taken to be the temperature of steepest descent in resistance, appears not to have been

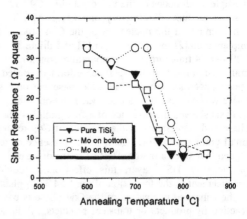

Fig. 1. Sheet resistance of TiSi$_2$ layers, grown on SiO$_2$, after 60s RTA in nitrogen.

lowered for either of the samples in contact with a $Ti_{0.4}Mo_{0.6}Si_2$ layer, in comparison with the pure "$TiSi_2$" film. Actually, the transformation for the layer with Mo on top has occurred at a higher temperature. A 15nm* thick, $Ti_{0.95}Mo_{0.05}Si_{2.0}$ layer, co-deposited on SiO_2 at room temperature, was also found to have a similar phase transformation temperature. TEM micrographs in Fig. 2 show the morphologies of the majority C49 layers as-grown at 550°C in UHV, before the C54 $TiSi_2$ transformation. All of these $TiSi_2$ layers, grown with or without the presence of molybdenum, have large grain sizes of over 200nm.

Fig. 2. Planview, bright-field, TEM images of Ti silicide layers as grown in UHV on SiO_2 at 550°C. The layers were grown by (a), "pure $TiSi_2$", the depositions of 15nm* $TiSi_{2.0}$ and 7.5nm Si at room temperature; (b), "Mo on bottom", the co-deposition of 2nm $Ti_{0.4}Mo_{0.6}Si_2$ at room temperature, an anneal at 630°C for 1min, and the co-deposition of 13.5nm $TiSi_{2.0}$ and 7.5nm Si at room temperature; and (c) "Mo on top", the depositions of 7.5nmSi and 13.5nm* $TiSi_{2.0}$ at room temperature, an anneal at 550°C for 1 min, and the deposition of 2nm $Ti_{0.4}Mo_{0.6}Si_2$ at room temperature.

$TiSi_{0.6}$ layers co-deposited on SiO_2 were found to remain amorphous up to 540°C. Crystalline Ti_5Si_3 phase formed at above 580°C. On top of a 5nm* thick Ti_5Si_3 layer grown at 650°C, a 10nm* thick $TiSi_{2.0}$ layer and a 10nm thick Si layer were deposited at room temperature and were given a mild 550°C anneal in situ. An alternative sample was grown using the same stack of layers except that the initial Ti_5Si_3 layer was not annealed, i.e. an amorphous template layer was used. The transformation of these two types of samples was observed to be similar to that for a co-deposited $TiSi_{2.0}$ layer, i.e. there appeared to be no enhancement in the formation of the C54-$TiSi_2$ phase due to Ti_5Si_3 templates. TEM analysis showed that the sample grown with a crystalline Ti_5Si_3 template went through the crystalline TiSi phase upon annealing prior to, or concurrently with, the formation of the C49-$TiSi_2$ phase, as shown in Fig. 3(b). The C49-$TiSi_2$ phase was the first crystalline phase detected for the sample with an amorphous Ti_5Si_3 template layer, as shown in Fig. 3(a). Note that the C49-$TiSi_2$ layer in Fig. 3(a) also had a large average grain size.

All the samples described so far involved the co-deposition of $CoSi_{2.0}$ at room temperature and they all had large grain size and relatively high transformation temperatures. To compare with previous results, layered structures were deposited on SiO_2 and their silicide reactions

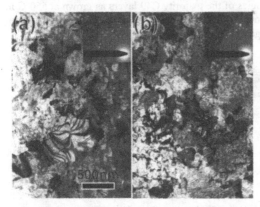

Fig. 3. Planview, bright-field, TEM images of 15nm* thick Ti silicide layers on SiO_2 at 550°C. The UHV growth conditions are (a) 5nm* $TiSi_{0.6}$ co-deposition at room temperature, 630°C 1min anneal, 10nm $TiSi_{2.0}$ co-deposition and 7.5nm Si deposition at room temperature. (b) 5nm* $TiSi_{0.6}$ co-deposition, 10nm $TiSi_{2.0}$ co-deposition, 7.5nm Si deposition, all at room temperature.

studied. Shown in Fig. 4 are TEM images of layered structures as-grown at 550°C in UHV. It is apparent that the $TiSi_2$ layer grown from deposited Ti/Si layered stack had a smaller grain size, < 100nm, than those grown from co-deposition at room temperature. With the addition of a 0.5nm Mo interlayer, the grain size is further reduced significantly, to ~20nm, and the transformation temperature was reduced by about 100°C, to ~620°C. These results are in good agreement with those obtained previously on single crystal or polycrystalline Si and they thus indicate an absence of contamination effects from the SiO_2 substrates presently used.

The grain size of the $TiSi_2$ layer has been artificially adjusted by using MBE and the template technique.[8] An ultrathin, 0.3-2nm*, $TiSi_{2.0}$ layer was co-deposited at room temperature on SiO_2 and annealed at 350-600°C , which led to the growth of $C49-TiSi_2$ layers with very variable grain sizes.   These ultra-thin layers were then used as template layers to grow $TiSi_2$ layers, with total thickness of 34nm, by co-deposition of $TiSi_{2.0}$ at ~250-350°C. The $C49-TiSi_2$ layers grown by this template technique have grain sizes which range from 25-

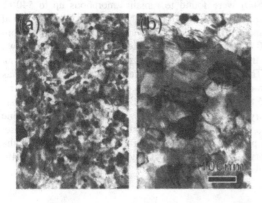

Fig. 4. Planview, bright-field, TEM images of 34nm TiSi2 layers grown on SiO2 by the depositions of (a) 45nm Si, 0.5nm Mo, and 15nm Ti, (b) 45nm Si and 15nm Ti, at room temperature and a 550°C 1min anneal in UHV.

70nm, depending on the template and the MBE growth conditions. An example of polycrystalline morphology of $TiSi_2$ layers grown by this technique is shown in Fig. 5(b). The transformation temperatures of these silicide layers are in the range of 640-700°C, considerably reduced from that described above for layers whose growth involved mainly co-deposition at room temperature. It was discovered that $TiSi_2$ layers grown with ultrathin $Ti_{0.4}Mo_{0.6}Si_2$ layers as the template layer, had very small grain sizes, typically ~15nm, under similar MBE conditions, as shown in Fig. 5(a). Quite fittingly, these layers were found have the lowest transformation temperature of all the samples studied.

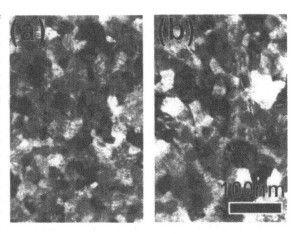

Fig. 5. Planview, bright-field, TEM images of examples of ~34nm $TiSi_2$ layers grown by low temperature MBE from (a) thin $Ti_{0.4}Mo_{0.6}Si_2$ template and (b) $TiSi_2$ template.

The transformation temperatures of all the Ti silicide samples studied in this work are plotted in Fig. 6 against the grain size of the $C49$-$TiSi_2$ phase, observed by TEM, prior to the transformation. A clear correlation of the transformation temperature with the grain size can be seen: the smaller the grain size, the lower the transformation temperature. Little distinction can be made of data for layers grown with and without Mo, as they appear to fall on the same curve. Such a behavior tends to de-emphasize possible chemical effects due to refractory metal impurities.[7] Also, the present data offer no support for template-related enhancement effect from either the C40-phase [3] or the Ti_5Si_3 phase.[6] The chemical and template effects, if they were present, seem secondary to the main role played by the refractory metal impurities: a reduction in the grain size of the Ti silicide under favorable conditions.

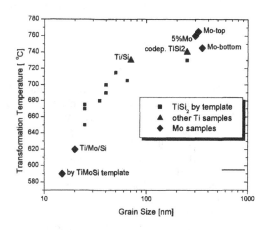

Fig. 6. Nominal transformation temperature of ~34nm $TiSi_2$ layers grown on SiO_2, against the TEM-observed average grain size of the as-grown (C49) $TiSi_2$ layer.

The grain size of a reacted thin film, in the absence of post-formation coarsening, is largely determined by the nucleation speed and the growth rate. In general, a higher nucleation rate and/or a slow growth rate leads to a smaller grain size. In the three cases in the literature where Rm impurities have been observed to enhance the formation of the C54 phase,[2-4] the locations of the Rm impurities have been such that the diffusion to form Ti silicide (e.g.C49) could be retarded. In each case the grain size has been dramatically reduced. In the present experiments involving the C40 $Ti_{0.4}Mo_{0.6}Si_2$ templates or the ternary $Ti_{0.95}Mo_{0.05}Si_{2.0}$ layer, the silicide growth is essentially a crystallization process from amorphous layers. Since no long-range diffusion is required, the growth speed is expected to be very large, resulting in large grain sizes regardless of whether Mo is present. The large grain size from deposited $Ti_{0.95}Mo_{0.05}Si_{2.0}$ layer is also suggestive of Mo incorporation in the silicide crystals, as opposed to significant segregation at grain boundaries. In a previous experiment, Mo implantation into pre-existing C49 $TiSi_2$ films also failed to produce an enhancement of the C54 $TiSi_2$ formation. One notices that all these cases where Rm showed no favorable influence on the $TiSi_2$ reaction were also cases where silicide growth required no solid-phase diffusion. These results are suggestive that how impurities such as Mo enhance the C54 TiSi2 formation is predominantly through a change in the (C49 $TiSi_2$) reaction kinetics. Energetic reasons appear to be of only secondary importance. The effectiveness of a metallic impurity thus likely depends on how it affects relevant kinetic processes such as Si diffusion and grain boundary segregation.

SUMMARY

UHV co-deposition technique has been used to study of the nucleation of $TiSi_2$ phase(s) in contact with a pre-existing C40 $Ti_{0.4}Mo_{0.6}Si_2$ or Ti_5Si_3 layer. No identifiable enhancement in the C54-$TiSi_2$ nucleation was observed which could be attributed to templates. Instead, the nucleation temperature of the C54-$TiSi_2$ phase appeared to be correlated with the grain size of the C49-$TiSi_2$ layer, independent of whether Mo was present. These results are suggestive that the primary mechanism for the enhanced formation of the C54 phase by refractory metals is a reduction in the grain size of the C49 $TiSi_2$ phase, likely due to altered kinetics.

REFERENCES

1. Z. Ma, L.H. Allen, and D.D.J. Allman, Thin Solid Films **253**, 451 (1994).
2. R.W. Mann, G.L. Miles, T.A. Knotts, D.W. Rakowski, L.A. Clevenger, J.M.E. Harper, F.M. D'Heurle, and C. Cabral Jr., Appl. Phys. Lett. **67**, 3729 (1995).
3. A. Mouroux, S.-L. Zhang, W. Kaplan, S. Nygren, M. Ostling, and C.S. Petersson, Appl. Phys. Lett. **69**, 975 (1996).
4. C. Cabral Jr., L.A. Clevenger, J.M.E. Harper, F.M. D'Heurle, R.A. Roy, C. Lavoie, K.L. Saenger, G.L. Miles, R.W. Mann, and J.S. Nakos, Appl. Phys. Lett. **71**, 3531 (1997).
5. S. Ohmi and R.T. Tung, J. Appl. Phys. **86**, 3655 (1999).
6. A. Quintero, M. Libera, C. Cabral Jr., C. Lavoie, and J.M.E. Harper, J. Mater. Res. **14**, 4690 (1999).
7. F. Bonoli, M. Iannuzzi, L. Miglio, and V. Meregalli, Appl. Phys. Lett. **73**, 1964 (1998).
8. R.T. Tung, J.M. Gibson, and J.M. Poate, Appl. Phys. Lett. **42**, 888 (1983).

Mat. Res. Soc. Symp. Vol. 611 © 2000 Materials Research Society

Investigation on C54 nucleation and growth by micro-Raman imaging

Stefania Privitera[1], Francesco Meinardi[2], Simona Quilici[2], Francesco La Via[3], Corrado Spinella[3], Maria Grazia Grimaldi and Emanuele Rimini[1]
1) INFM and Physics Department, Catania University,
Corso Italia 57, 95129 Catania, Italy
2) INFM and Materials Science Department, Milano-Bicocca University,
Via Cozzi 53, 20125 Milano, Italy
3) CNR-IMETEM,
Stradale Primosole 50, 95121Catania, Italy

ABSTRACT

The processes of nucleation and growth of the C54 $TiSi_2$ phase into the C49 phase in thin films have been studied by electrical measurements and micro-Raman spectroscopy. The Raman spectra have been acquired scanning large silicide areas (100×50 μm^2) in step of 0.5 μm. Images showing the evolution of the C54 grains during the transition have been obtained for temperatures between 680 and 720 °C and the transformed fraction, the density and the size distribution of the C54 grains have been measured as a function of the temperature and the annealing time. The activation energies for the nucleation rate and the growth velocity have been determined obtaining values of 4.9 ± 0.7 eV and 4.5 ± 0.9 eV, respectively.

INTRODUCTION

The formation of the low resistivity C54 phase of $TiSi_2$ is of great importance for the implementation of Ti salicide process into deep submicron devices[1]. In thin film reactions, however, the C54 phase is always preceded by the metastable high resistivity C49 phase and a high temperature anneal (> 700 °C) is required to transform the silicide into the technologically required phase.

The C49-C54 allotropic phase transition has been largely investigated by several techniques but the processes of nucleation and growth of the C54 are still not clear. To understand and control these processes a detailed description of the driving force for the transformation and the nucleation barrier energy is required. The link between the kinetics of the transition and the driving force is the dynamical evolution of the population of C54 grains during the transformation. However, this information is not accessible by the most commonly used techniques, such as electrical measurements and X-ray diffraction, which are only sensitive to the transformed volume fraction. In this work micro-Raman spectroscopy, which allows the silicide phase to be locally identified [2-4], has been used to obtain images showing the morphological evolution of C54 grains into the C49 phase. The transformed fraction, the density of grains and the size distribution have been determined for various temperatures and times and have been interpreted in terms of the classical nucleation and growth theory.

EXPERIMENTAL PROCEDURES

A 300 nm undoped amorphous silicon layer was grown by low-pressure chemical vapour deposition (LPCDV) on thermally oxidised Si (100) wafers. Immediately before the deposition the substrate was cleaned in HF to remove the native oxide and a 15 nm Ti layer has been deposited in a e-gun evaporator at a pressure of 10^{-9} Torr. The C49 silicide phase was formed in N_2 by a rapid thermal anneal (RTA) at 650 °C for 5 min. The unreacted Ti was removed in a $NH_4OH:H_2O_2:H_2O$ (1:1:5) solution and the film was converted into the C54 by rapid thermal anneals at 680, 700 and 720 °C. Anneals were interrupted at different times and the transition was monitored ex-situ by electrical measurements and micro-Raman spectroscopy. The thickness and the composition of the silicide were determined by Rutherford backscattering spectrometry (RBS) of 2 MeV $^4He^+$ ions in grazing angle and by transmission electron diffraction. The sheet resistance has been measured using a four-point probe and the Raman measurements have been carried out with a Labram (Dilor) confocal spectrometer in backscattering configuration equipped with a 17 mW HeNe laser. The laser spot was focused into a 0.7 µm diameter spot and spectra were collected moving the sample in step of 0.5 µm. The total analyzed area for each sample was 100 x 50 µm² and the integration time for each spectrum was 4 s.

RESULTS

The Raman spectrum of the C54 phase is characterised by a peaks envelope around 240-250 cm^{-1}, detectable regardless the crystal orientation respect to the laser probe beam [5], while the C49 phase is identified [6] by three Raman peaks at 270, 330 and 355 cm^{-1}. Thus the two phases can be easily distinguished. To determine the C54 fraction in each tested point we considered the intensity of both the C54 and C49 Raman signals. The minimum detectable C54 area depends on the signal to noise ratio and in our experimental conditions the minimum detectable C54 grain radius was $r_d = 0.1$ µm. Figure 1 (a) and (b) reports typical micro-Raman images (MRI) of partially transformed samples annealed at 680 °C for 150 and 210 s, respectively. The bright spots represent the C54 phase surrounded by the C49 grains (black regions). The number and the size of the C54 grains increase with the annealing time, indicating that the transition proceeds via nucleation and growth.

Figure 2 shows the transformed fraction (a) and the C54 grain density (b) as a function of annealing time. In Fig. 2(a) we compared the C54 fraction obtained from micro-Raman images (full symbols) with the fraction X(t) obtained from electrical measurements using the relationship $X(t) = (R_{C49}-R(t))/ (R_{C49} - R_{C54})$, where R_{C49} and R_{C54} are the sheet resistance of the C49 and the C54 phase, respectively and R(t) is the sheet resistance measured at the time t [7,8].

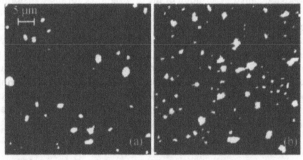

Figure 1. *Micro-Raman imaging of samples annealed at 680 °C for 150 s (a) and 210 s (b). White regions are C54 grains inside the C49 phase (black area).*

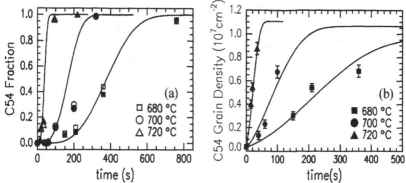

Figure 2. *Transformed fraction (a) and C54 grain density (b) v.s. annealing time. Data obtained from electrical measurements are showed by open symbols. Filled symbols are data obtained from MRI. Squares, circles and triangles represent samples annealed at 680, 700 and 720 °C, respectively. The continuous lines are fits to data.*

The comparison shows a quite good agreement between the two methods. Squares, circles and triangles indicate samples annealed at 680, 700 and 720 °C, respectively.

Moreover, the grain density showed in Fig. 2 (b) increases with the time and, as the temperature increases, the average nucleation rate J, i.e. the number of C54 grains that appear per unit time and area, exponentially increases.

In Fig. 3 (a), (b) and (c) the grain size distributions (GSD) of samples annealed at 680 °C for 150 s, 700 °C for 40 s and 720 °C for 20 s, respectively, have been reported. For detectable C54 grain sizes, the GSD n(r,t), is given by the equation [9]

$$n(r,t) = \frac{J(r,t)}{v} \qquad (1)$$

where v is the growth velocity and the nucleation rate J(r,t) is defined as the density of nuclei which grow above the dimension r per unit time. Assuming that the processes of nucleation and growth proceed only in two dimensions, the transformed fraction as a function of annealing time, is given by the following relationship

$$\chi(t) = 1 - \exp(-\chi_{ext}) \qquad (2)$$

where $\chi_{ext} = \int_0^t \pi \varepsilon v^2 (t-t')^2 J(\varepsilon/2, t') dt'$ is the extended volume fraction of the C54 at the time t

and ε is the film thickness [10]. The C54 grain density $\rho(t)$ measured at the time t depends on the minimum detectable size and on the volume available for the nucleation. Therefore, $\rho(t)$ is given by the expression

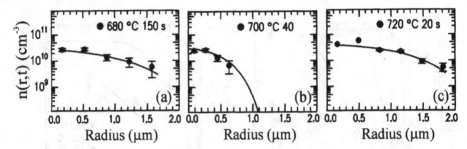

Figure 3. *Grain size distributions measured in samples annealed at 680 °C for 150s (a), 700 °C 40 s (b), and 720 °C 20 s (c). The bars are statistical errors. The continuous lines are fits to the data.*

$$\rho(t) = \varepsilon \int_0^t [1 - \chi(t')] J(r_d, t') dt' \qquad (3)$$

Equations (1), (2) and (3) have been used to simultaneously fit the grain size distribution, the transformed fraction and the grain density, respectively. For the nucleation rate $J(r,t)$ we used the analytical expression proposed by Shneidmann [11,12] :

$$J(r,t) = J_S \exp\left[-\exp\left(\frac{t_0 + r/v - t}{\tau_s}\right)\right] \qquad (4)$$

The incubation time t_0 is the time spent by a nucleus to grow from zero to a size $i^* - \delta/\sqrt{2}$, being δ the half width of the region where the cluster free energy differs less than KT from the maximum value ΔG^* and i^* is the critical radius. The transient time τ_s is the time during which the size of the nucleus is $i^* - \delta/\sqrt{2} < i < i^* + \delta/\sqrt{2}$. The parameters that we used in the fitting

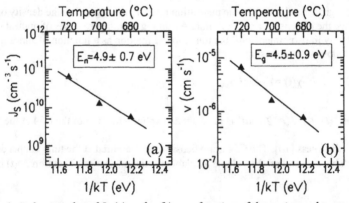

Figure 4. *Arrhenius plot of J_S (a) and v (b) as a function of the reciprocal temperature.*

procedure are those appearing in Eq. (4). The comparison between experimental and theoretical data has been showed in Fig. 2 and 3 as solid lines. Figure 4 (a) and (b) shows the Arrhenius plot of J_S and v. The obtained activation energies are $E_n = 4.9 \pm 0.7 eV$ and $E_g = 4.4 \pm 0.8$ eV, respectively.

DISCUSSION

The activation energies determined for the transient and incubation times are coincident, within uncertainty, and equal to the activation energy for the growth rate, as we can expect since t_0 and τ_s both depend on the process of growth.

The value of E_g that we determined is in very good agreement with data reported in literature evaluated either by TEM analyses in situ (4.6 eV) [13] or electrical measurements (4.4 eV) by fitting the Avrami-Mehl-Johnson equation [10] to the transformed fraction, with an exponent n = 2, which indicates two dimensional growth and nucleation rate equal to zero [7,8].

The activation energy for the nucleation, in the classical theory of nucleation and growth, is defined as the sum of the activation energy for the growth velocity E_g plus the barrier energy for the nucleation ΔG^*, i.e. $E_n = E_g + \Delta G^*$. Therefore we should be able to evaluate the barrier for the nucleation. However, the value of E_n that we found is equal, within experimental errors, to E_g and this implies that ΔG^* is extremely low. Such a result could be explained by considering heterogeneous nucleation. Indeed, in this case ΔG^* is lower than the barrier energy for homogeneous nucleation for a factor $f(\theta)$ which depends on the geometrical configuration of the nucleation sites and, for particular values of the contact angle θ, ΔG^* can approach to zero [10].

This, in agreement with previous observations [13,14], means that only few nucleation sites, which satisfy the condition $\Delta G^* \rightarrow 0$, are active.

CONCLUSION

Micro-Raman imaging have been used to determine the converted fraction, the C54 grain density and size distribution as a function of annealing times for temperatures between 680 and 720 °C. The contribution of nucleation and growth to the phase transition has been separated and the parameters that characterise the kinetics of the transformation have been determined. The activation energy for the growth rate and the nucleation rate has been evaluated. Values coincident within errors have been found, indicating that the barrier energy for the nucleation is very low.

REFERENCES

1. K. Maex, Mater. Sci, Eng., R **11** (1993) 53.
2. F. Meinardi, L. Moro, Sabbadini and Queirolo, *Europhys. Lett.*, **44**, 57 (1998)
3. DeWolf I. De Wolf, D.J. Howard, A. Lawers, K. Maex and H. E. Maes Appl. Phys. Lett. **70** (1997) 2262
4. H. E. Lim, G. Karunasiri, S. J. Chua, H. Wong, H. L. Pey and K. H. Lee *IEEE Elect. Dev. Lett.*, **19** (1998) 171
5. F. Meinardi, S. Quilici, A. Borghesi, G. Artioli Appl. Phys. Lett. **75** (1999) 3090.

6. H. Jeon, C.A. Sukow, J. W. Honeycutt, G. A. Rozgonyi and R. J. Nemanich, J. Appl. Phys. **71** (1992) 4269.
7. Z. Ma and L.H. Allen, *Phys. Rev. B*, **49**, 13501 (1994).
8. Z. Ma, L.H. Allen and D. D. J. Allman, *J. Appl. Phys.*, **77**, 4384 (1995).
9. R. C. Spinella, S. Lombardo, and F. Priolo, *J. Appl. Phys.*, **84**, 5383 (1998).
10. J. W. Christian, *The theory of Transformation in Metals and Alloys*, Part. I 2nd ed., Pergamon, (Oxford, 1995).
11. V. A. Shneidman, *Sov. Phys. Tech. Phys.*, **32**, 76 (1987).
12. V. A. Shneidman, *Sov. Phys. Tech. Phys.*, **33**, 1338 (1988).
13. L. M. Gignac, V. Svilan, L. A. Clevenger, C. Cabral, Jr. and C. Lavoie, *MRS Proceed.*, **441**, 255 (1997).
14. S. Privitera, F. La Via, M. G. Grimaldi and E. Rimini, *Appl. Phys. Lett.*, **73**, 3863 (1998).

Mat. Res. Soc. Symp. Vol. 611 © 2000 Materials Research Society

Silicide engineering : influence of alloying elements on CoSi$_2$ nucleation

C. Detavernier[*], R.L. Van Meirhaeghe[*], K. Maex[+0], F.Cardon[*]

[*] Vakgroep Vaste Stofwetenschappen, Ghent University, Krijgslaan 281/S1, 9000 Gent, Belgium
[+] IMEC, Kapeldreef 75, B-3001 Leuven, Belgium.
[0] also at E.E. Dept, K.U. Leuven, B-3001 Leuven, Belgium.

ABSTRACT

Evidence is presented that the nucleation of CoSi$_2$ can be influenced by the presence of small amounts of other elements. The presence of trace amounts of Ti in the CoSi (originating from either a Ti capping layer or interlayer) causes an increase in the CoSi$_2$ nucleation temperature. Moreover, the presence of Ti in the CoSi induces a preferential orientation of the CoSi$_2$: for an increasing amount of Ti, we observed a transition from polycrystalline CoSi$_2$ over preferential (220) orientation towards epitaxial (400) CoSi$_2$. We observed similar effects for other elements (e.g. Ta, W, C, Mo, Cr). We were able to explain these findings based on the heterogeneous nucleation of CoSi$_2$.

INTRODUCTION

The presence of a 3-7 nm Ti interlayer (in between the Co and Si substrate) is known to promote epitaxial growth of CoSi$_2$ (Ti Interlayer Mediated Epitaxy, TIME)[1]. Traditionally, this is explained by (1) the ability of Ti to reduce SiO$_2$ and by the fact that (2) the interlayer acts as a diffusion barrier. In this work, we studied phase formation and preferential orientation of the CoSi$_2$ as a function of Ti thickness, both for interlayers (0.1-10 nm) and capping layers (1-10 nm). In case of a thin Ti interlayer (< 1 nm), we observed that the nucleation of CoSi$_2$ is delayed to about 600°C. Moreover, the CoSi$_2$ has a strong preferential orientation and improved thermal stability. For 0.1-0.6 nm Ti, there is a (220) and (400) orientation, while for thicker Ti interlayers (> 1 nm), there is a strong <400> orientation. Similar results are found for a Ti capping layer (Ti/Co/Si system).
Our results indicate that there are two regimes of interlayer thickness wherein the epitaxial growth is achieved by two different mechanisms : for thick interlayers, the function of the interlayer is to slow down diffusion, while for thin interlayers (< 1 nm), the main function of Ti is to improve the grain boundary cohesion in the CoSi. We present a model on heterogeneous nucleation of CoSi$_2$, which explains how even a trace amount of Ti is able to strongly influence the nucleation of CoSi$_2$, because of its influence on interfacial energy.

EXPERIMENT

For all experiments, p-type Si(100) substrates were used. After cleaning and a final HF dip, the substrates were mounted in an e-beam evaporation system. Co and Ti layers were deposited in a vacuum of 10^{-6} mbar. Bilayers were deposited without breaking the vacuum. Both the effect of Ti as an interlayer (Co/Ti/Si) and capping layer (Ti/Co/Si) was investigated. To study the silicidation reaction, isochronal RTP annealing (30s, N_2 ambient) was done at various temperatures. Sheet resistance measurements and grazing incidence X-Ray Diffraction (XRD) were used for phase identification. Standard $\theta/2\theta$ XRD measurements and Rutherford Backscattering Spectroscopy (RBS) were used to determine preferential orientation and epitaxial quality of the $CoSi_2$ layers.

RESULTS

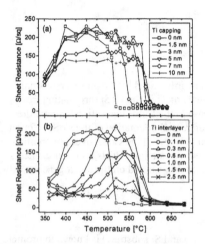

Figure 1 : Sheet resistance versus annealing temperature for a Ti/Co(10 nm)/Si system (a) and a Co(10 nm)/Ti/Si system (b).

Figure 2 : XRD intensity for the $CoSi_2$ (111), (220) and (400) peaks versus Ti capping layer thickness (a) and interlayer thickness (b) for samples annealed at 800°C. For clarity, the intensities were rescaled using the scaling factors indicated in the legend.

In figure 1a, the sheet resistance is plotted versus the annealing temperature for different values of the Ti capping layer thickness. It is clear that the formation of the low resistive $CoSi_2$ phase is gradually delayed from about 520°C for a standard Co/Si system to about 580-600°C for a Ti(10nm)/Co/Si system.

In figure 1b, the sheet resistance is plotted in case of a Ti interlayer. For 0.1 nm Ti, CoSi formation is not influenced significantly as compared to the reference without Ti interlayer, while $CoSi_2$ formation is delayed from 520°C to about 560-580°C. For slightly thicker Ti interlayers (0.3 – 1 nm), both CoSi and $CoSi_2$ formation are delayed. The delay in CoSi formation may be explained by the fact that Co has to diffuse through the interlayer before it can react with the Si substrate. Nucleation of $CoSi_2$ is delayed up to about 580°C. For thicker Ti interlayers (> 1nm), we approach the regime of traditional TIME.

We studied the preferential orientation of the $CoSi_2$ layer by plotting the intensity of the various $CoSi_2$ XRD peaks versus the Ti layer thickness (figure 2b). In case of a Ti interlayer, it is clear that small amounts of Ti strongly influence the preferential orientation of the $CoSi_2$. For the

Co/Si reaction without Ti, a polycrystalline CoSi₂ layer is formed, and in XRD the (111) and (220) peaks are present. For small amounts of Ti (0.1 nm interlayer), the (111) peak intensity decreases, while the (220) intensity increases. For a further increase in thickness of the Ti interlayer, the (111) peak disappears, the (220) intensity decreases again and the (400) intensity increases sharply. In RBS measurements, already for a very thin Ti interlayer (0.1 nm), some channeling effect is observed. It appears that due to the presence of Ti, there is a selective nucleation of substrate-matched nuclei. In case of a Ti capping layer, the preferential orientation versus capping layer thickness shows qualitatively the same behavior (figure 2a), although a thicker capping layer is needed. Qualitatively, it seems that the behavior for Ti

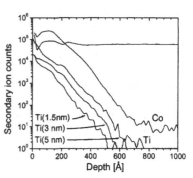

Figure 3 : SIMS profile for a Ti/Co(10 nm)/Si structure with a Ti capping thickness of 1.5, 3 and 5 nm, annealed at 520°C (after selective etching in H SO).

interlayers with a Ti thickness between 0 and 1 nm is reproduced for a capping thickness between 0 and 10 nm. This is caused by the consumption of Ti from the capping layer by reaction with the N_2 annealing ambient.

For the Ti capping layer, SIMS measurements indicated that Ti from the capping layer diffuses into the CoSi (figure 3). The thicker the capping layer, the more Ti is present in the CoSi. For the interlayer, since the Ti is deposited in between the Co and Si, it is clear that Ti will also be present in the CoSi layer.

GENERAL MODEL : INFLUENCE OF ALLOYING ELEMENTS ON COSI₂ NUCLEATION

It has been reported that the CoSi → CoSi₂ transition is nucleation controlled, with diffusion as a rate limiting process during the growth of the nuclei. According to the classical theory of nucleation[2], the activation energy for nucleation is given by $\Delta G^* \approx \Delta\sigma^3 / \Delta G^2$, with ΔG the change in free

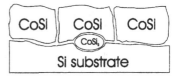

energy and Δσ the change in interfacial energy induced by the formation of the nucleus. Since the CoSi is polycrystalline, there are two different types of locations where the CoSi₂ may nucleate : in the middle of a CoSi grain (homogeneous nucleation) or at a CoSi/CoSi grain boundary (heterogeneous nucleation). In the latter case, the grain boundary energy σ{CoSi / CoSi} > 0 will result in a lower activation energy for nucleation, since

$$\Delta\sigma_{heterogeneous} = \sigma\{CoSi_2/Si\} + \sigma\{CoSi_2/CoSi\}$$
$$-\sigma\{CoSi/Si\} - \sigma\{CoSi/CoSi\}$$

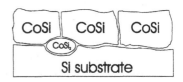

From XTEM measurements, it has been shown that the nucleation of CoSi₂ is indeed initiated at the grain boundaries in the CoSi precursor phase [3].

Figure 4 : Homogeneous versus hererogeneous nucleation.

Nucleation temperature and preferential orientation

In case of nucleation of CoSi$_2$ from CoSi, one has to consider the following activation energy :

$$\Delta G^{*} \approx \frac{\left(\sigma_{CoSi_2/Si} + \sigma_{CoSi_2/CoSi} - \sigma_{CoSi/Si} - \sigma_{CoSi/CoSi}\right)^{3}}{\left(\Delta H - T\Delta S + \Delta H_{strain}\right)^{2}} + Q \tag{1}$$

Factors influencing the activation energy for nucleation of the CoSi$_2$ are (1) the change in formation enthalpy ΔH, (2) the entropy change ΔS, (3) surface energy terms and (4) a kinetic term Q. The kinetic term takes into account the activation energy needed for the local atomic rearrangement needed to form the nucleus. This rearrangement will most likely be based on diffusion. The Q mentioned here is however not necessarily the same activation energy as needed for the growth (thickening) of the CoSi$_2$ layer. Calculations by S.L. Zhang have shown that in the case of CoSi$_2$ the strain energy ΔH_{strain} can be estimated to be about ten times smaller than ΔH [4].

Because of the presence of $\sigma\{CoSi_2/Si\}$ in this formula, the activation energy for nucleation will depend on the orientation of the CoSi$_2$ nucleus with respect to the Si substrate :

$$\Delta G_{400}^{*} \approx \frac{\left(\left[\sigma_{CoSi_2/Si}\right]_{400} + \sigma_{CoSi_2/CoSi} - \sigma_{CoSi/Si} - \sigma_{CoSi/CoSi}\right)^{3}}{\left(\Delta H - T\Delta S + \left[\Delta H_{strain}\right]_{400}\right)^{2}} + Q$$

$$\Delta G_{220}^{*} \approx \frac{\left(\left[\sigma_{CoSi_2/Si}\right]_{220} + \sigma_{CoSi_2/CoSi} - \sigma_{CoSi/Si} - \sigma_{CoSi/CoSi}\right)^{3}}{\left(\Delta H - T\Delta S + \left[\Delta H_{strain}\right]_{220}\right)^{2}} + Q$$

$$\Delta G_{111}^{*} \approx \frac{\left(\left[\sigma_{CoSi_2/Si}\right]_{111} + \sigma_{CoSi_2/CoSi} - \sigma_{CoSi/Si} - \sigma_{CoSi/CoSi}\right)^{3}}{\left(\Delta H - T\Delta S + \left[\Delta H_{strain}\right]_{111}\right)^{2}} + Q$$

Based on previous work by Bulle-Lieuwma, one can expect that $\sigma\{CoSi_2/Si\}_{400} < \sigma\{CoSi_2/Si\}_{220} < \sigma\{CoSi_2/Si\}_{111}$ [5], thus $\Delta G_{400}^{*} < \Delta G_{220}^{*} < \Delta G_{111}^{*}$.

The various factors that determine the nucleation can be controlled :

1. Nucleation will only be rate limiting if ΔH is small. This is the case if CoSi$_2$ is formed from CoSi, because the difference in heat of formation is very small between these two phases. If one would be able to form CoSi$_2$ in a direct reaction between Co and Si, the ΔH for this reaction would be much larger, resulting in a very low activation energy for nucleation. MBE experiments have shown that this is possible if there is a very slow supply of Co atoms towards the Si substrate [6]. In solid state reactions, the same effect may be obtained by adding a diffusion mediating layer in between the Co layer and the silicon substrate. In the case of a thin SiO$_2$ interlayer, it has been reported that CoSi$_2$ is the first phase to form [7]. Another way to influence ΔH is to use amorphous Si as a substrate. In this case there is an additional crystallization energy for Si. This causes ΔH to increase. Indeed, it has been found experimentally that the CoSi$_2$ formation on a-Si is no longer nucleation controlled [2].

2. The entropy change ΔS may be influenced by adding 'foreign elements' to the reaction that have a different solubility in CoSi and CoSi$_2$, inducing a difference in entropy of mixing [8].

3. The surface energy terms may be influenced by adding small amounts of impurities that are insoluble in CoSi or CoSi$_2$. In this case, the impurities will be present at the grain boundaries and may influence the grain boundary cohesion $\sigma\{CoSi/CoSi\}$.

4. The kinetic term may also be influenced by adding impurities to the grain boundaries. This may influence grain boundary diffusion.

In this way, one can control both the $CoSi_2$ formation temperature and preferential orientation :
- the **formation temperature** will be dependent on the magnitude of ΔG^* : to decrease the formation temperature, one will have to decrease the activation energy for nucleation.
- the **preferential orientation** is determined by $\Delta(\Delta G^*)$: as mentioned before, there will be a difference in activation energy for the different possible orientations of the $CoSi_2$ with respect to the Si substrate. The difference in these activation energies $\Delta(\Delta G^*)$ will determine the amount of preferential orientation : if there is a large difference between ΔG_{400}^* and ΔG_{111}^*, the layer will nucleate epitaxially.

In the standard Co/Si reaction, contrary to what one may expect, the $CoSi_2$ is polycrystalline. We may conclude from this that for the standard reaction $\Delta(\Delta G^*)$ is small. This may explained if $\Delta\sigma\{CoSi_2/Si\}$ (the difference in interfacial energy for the different orientations of $CoSi_2$) is small compared to the grain boundary energies $\sigma\{CoSi_2/CoSi\}$ and $\sigma\{CoSi/CoSi\}$. In this case the latter terms will always dominate ΔG^*, in spite of small differences in $\sigma\{CoSi_2/Si\}$ for the different orientations of the $CoSi_2$ nucleus. Moreover, because of the lattice mismatch, the formation of an epitaxial nucleus causes intrinsic stress, thus $[\Delta H_{strain}]_{400} >> [\Delta H_{strain}]_{111}$. According to eqn. (1), the preferential orientation of the $CoSi_2$ can be controlled in several ways :
1. If one can lower the grain boundary energy $\sigma\{CoSi/CoSi\}$, then $\sigma\{CoSi_2/Si\}$ will become more important, and $\Delta(\Delta G^*)$ will increase, resulting in a preferential epitaxial orientation.
2. A similar effect will occur if $\sigma\{CoSi/Si\}$ is lowered (e.g. if the CoSi layer has a preferential orientation or an improved cohesion with the substrate).
3. If one can increase ΔH (by bypassing the CoSi phase), nucleation will no longer be the rate limiting process. In this case, $CoSi_2$ will be the first phase to form and in order to minimise the interfacial energy $\sigma_{CoSi_2/Si}$ the layer will grow epitaxially on the Si substrate. This explains the epitaxial growth in MBE experiments and in case of diffusion mediating interlayers [6].

DISCUSSION

For thin Ti interlayers and Ti capping layers, we observed that CoSi is the first phase to form. Moreover, from the SIMS data, no evidence was found that Ti from the capping layer piles up at the CoSi/Si interface. Thus, in the case of a Ti cap or a very thin Ti interlayer, one should consider the Ti as a 'contaminant' present at the reaction front, rather than as a diffusion mediating interlayer. Our SIMS data indicate that Ti is present in the CoSi, where it piles up on the grain boundaries and may slow down grain boundary diffusion, increasing Q. Moreover, due to its reactive nature, Ti-Si bonds will be formed [9]. This will strengthen the grain boundary and thus lower the grain boundary energy $\sigma\{CoSi/CoSi\}$. This will result in an increase of $\Delta\sigma$, thus increasing ΔG^* (eqn.1). Moreover, the difference $\Delta(\Delta G^*)$ will increase, inducing a preferential orientation of the $CoSi_2$. Thus, in the presence of impurities at the CoSi grain boundary that enhance the grain boundary cohesion, heterogeneous nucleation theory predicts an increase of the nucleation temperature and a natural selection of the substrate matched nuclei ('survival of the best fitted'). On the other hand, in case of impurities that cause grain boundary decohesion, the theory predicts a decrease in the nucleation temperature. We have found that ultra thin interlayers of Ta, W, Cr, Mo and C show similar effects as Ti, delaying $CoSi_2$ nucleation to higher temperature and inducing epitaxial growth. In the presence of sulfur (20 nm of Co was evaporated in H_2S ambient), which is known to cause grain boundary decohesion in metals, we observed that $CoSi_2$ forms at the same or slightly lower temperature as compared to the standard Co/Si reaction.

Based on the results mentioned above, it seems that the TIME process works because of two different mechanisms : for very thin Ti layers, Ti should be considered a 'contaminant' in the CoSi, which is the first phase to form. The presence of Ti on the CoSi grain boundaries will influence the nucleation of $CoSi_2$ and will result in selective nucleation of substrate-matched nuclei. For thick Ti interlayers, the interlayer acts as a diffusion barrier, causing a small flux of Co towards the substrate.

SUMMARY

We presented a model for the heterogeneous nucleation of $CoSi_2$ and the influence of foreign elements on the nucleation reaction. This model allowed us to explain the increase in nucleation temperature and the preferential orientation of the $CoSi_2$ that was observed when small amounts of Ti are present in the CoSi. It is shown that Ti changes the nucleation reaction because its presence decreases the CoSi grain boundary energy. In an accompaying paper, the influence of the entropy term ΔS is shown for the Co-Fe-Si, Co-Ge-Si and Co-Ni-Si system [8].
These results illustrate the possibility of "silicide engineering" : the addition of small amounts of impurities (as a capping layer, interlayer or alloyed within the Co layer) may be used to control the silicidation reaction (reaction temperature), and to influence the properties of the silicide layer formed (preferential orientation, thermal stability).

ACKNOWLEDGEMENTS

The authors would like to thank Ing. L. Van Meirhaeghe for technical support. C. Detavernier thanks the 'Fonds voor Wetenschappelijk Onderzoek – Vlaanderen' (FWO) for a scholarship. K. Maex is a research director for the FWO.

REFERENCES

1. M. Lawrence, A. Dass, D.B. Fraser, and C.S. Wei, Appl. Phys. Lett. **58,** 1308 (1991).
2. F.M. d'Heurle, J. Mater. Res. **3,** 167-195 (1988).
3. A. Appelbaum, R.V. Knoell, S.P. Murarka, J. Appl. Phys. **57,** 1880 (1985).
4. S.-L. Zhang, F.M. d'Heurle, summerschool on silicides, Erice (1999).
5. C. Bulle-Lieuwma, A. van Ommen, J. Hornstra, C. Aussems, J. Appl. Phys. **71,** 2211 (1992).
6. A. Vantomme, S. Degroote, J. Dekoster, G. Langouche, R. Pretorius, Appl. Phys. Lett. **74,** 3137 (1999).
7. M.W. Kleinschmit, M. Yeadon, J.M. Gibson, Appl. Phys. Lett. **75,** 3288 (1999).
8. C. Detavernier, R.L. Van Meirhaeghe, F. Cardon, K. Maex, MRS 2000 (also in this volume).
9. T. Komeda, T. Hirano, G.D. Waddill, S.G. Anderson, J.P. Sullivan, J.H. Weaver, Phys. Rev. B **41,** 8345 (1990).

Shallow Junctions and Silicides

Mat. Res. Soc. Symp. Vol. 611 © 2000 Materials Research Society

Study of CoSix Spike Leakage for 0.1-um CMOS

Ken-ichi Goto
Fujitsu Laboratories Ltd., 10-1 Morinosato-Wakamiya, Atsugi 243-0197 Japan

ABSTRACT

We have clarified a new leakage mechanism in Co salicide process for the ultra-shallow junctions of 0.1-um CMOS devices and revealed the optimum Co salicide process conditions for minimizing the leakage current. We found that leakage currents generate from many localized points that are randomly distributed in the junction area, and not from the junction edge. We successfully verified our localized leakage model using Monte Carlo simulation. We identified abnormal CoSix spiking growth under the Co silicide film, as being the origin of the localized leakage current. These CoSix spikes grow rapidly only during annealing between 400°C and 450°C when Co2Si phase is formed. These spikes never grow during annealing at over 500°C, and decrease with high temperature annealing over 500°C. A minimum leakage current can be achieved by optimized annealing at between 800°C and 850°C for 30 sec. This is because a trade-off between reducing the CoSix spikes and preventing the Co atom diffusion from Co silicide film to Si substrate, which begins at annealing above 900°C.

INTRODUCTION

In deep sub-micron complementary metal oxide semiconductor (CMOS) devices, reduction of parasitic resistance is a key issue for high-speed operations [1]. A self-aligned silicide (salicide) process is an attractive approach for reducing both the sheet resistance of the gate and source/drain regions and the contact resistance to them.

Recently, Ti salicide process has been widely used in mass production of LSI logic devices. However, with shortening of gate length, it becomes more difficult to achieve low sheet resistance in the Ti salicide process by the following two reasons. One is a difficulty in a crystal phase transformation from high resistivity C49 phase to low resistivity C54 phase, and the other is agglomeration of the Ti silicide film in gate structures shorter than 0.1-um long [2], [3].

As an alternative material, Co salicide process has been intensively used. Recent studies have reported that by using a TiN-capping process, Co salicide successfully realizes the low gate sheet resistance of 5 ohm/sq down to 0.075-um gate length [1, 2, 4]. However, the Co salicide process causes a severe junction leakage, which becomes increasingly serious when the junction depth was scaled down to 0.1-um and below.

This paper reviewed the junction leakage mechanism of Co salicide process for 0.1-um CMOS technology. We verified this leakage model using Monte Carlo simulation, and identified CoSi₂ spikes which grow abnormally under the Co silicide film as the origin of the leakage current. We developed a low-leakage and low-resistance Co salicide process by using a high temperature annealing and a TiN-capping process for sub-0.1-um CMOS devices.

EXPERIMENTS

The n+/p-diodes were fabricated with a conventional LOCOS process and implanted with boron ions under the same conditions as in n-MOSFET channel fabrication. Then, a 0.1-um-deep ultra-shallow n/p-junction was formed by As+ implantation at 40 keV followed by rapid thermal annealing (RTA) in N2 at 1000°C for 10 sec. After HF treatment, we sputtered 10 nm-thick Co and 30 nm-thick TiN films. This Co thickness forms 35–40 nm-thick CoSi2 film with a sheet resistance of about 5ohm/sq. To analyze the leakage mechanism we varied the first RTA temperatures from 400 to 550°C for 30 sec to form Co2Si and/or CoSi. The TiN and un-reacted Co films were then chemically removed using sulfuric acid. Next, fixing the first RTA temperature at 550°C, a second RTA at various temperatures ranging from 750 to 900°C for 30 sec were performed. Circular diodes with radiuses of 50, 100, and 300 um were prepared to study area dependence, and square diodes with various peripheral lengths from 2.5 to 164 mm to study peripheral dependence. For statistical analysis, we measured 25 or 50 junctions fabricated under each condition at a reverse bias voltage of 2.5 V.

RESULTS AND DISCUSSION

Electrical Analysis for Localized Leakage Current

Co salicided junction characteristics always exhibit no repeatability among chips and wafers therefore statistical analysis is very effective way to understand the leakage mechanism. Fig. 1 show the reverse bias characteristics of Co salicided junctions after the second annealing at 750°C for 30 sec which is generally used because it provides sufficient temperature to reduce the sheet resistance. In this study, twenty-five chips of 320-um square diodes were measured. The typical leakage current sharply increased for the bias voltage up to about 5 V, and these currents became saturated beyond this voltage. And also, they varied widely from 1E-10 to 1E-6 A/diode at 2.5 V. These un-repeatable results made it difficult to understand the leakage mechanism.

In order to understand the mechanism, key experiments were done as follows [5]. The cumulative probabilities of the leakage current were measured in diodes of various peripheral length and areas, as shown in Figs. 2 and 3. These data revealed that the leakage current is independent of the

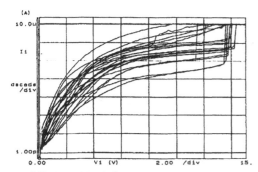

Fig. 1. *I-V* characteristic of n⁺p junction after Co salicide process with second annealing at 750 °C for 30 s.

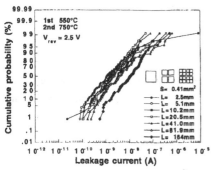

Fig. 2. Cumulative probability of leakage current at reverse bias voltage of 2.5 V with peripheral lengths from 2.5 to 164 mm.

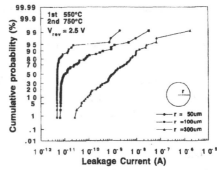

Fig. 3. Cumulative probability of leakage current at reverse bias voltage of 2.5 V with junction areas having radii of 50, 100, and 300 μm.

junction periphery, but extremely depend on the junction area. That is, the leakage current does not flow at the junction periphery but the junction area. Furthermore, the area dependence gives us more important information as follows. All the large junctions with a 300-um radius showed widely distributed leakage currents. However, by scaling the junction size down to a radius of 100-um, about 50% of the junctions retained as the same leakage level as that without salicide. And moreover, in the case of 50-um radius junction, over 80% retained at the low leakage level. These results suggest that the leakage current does not flow uniformly over the entire junction area, but through many localized points, which are randomly distributed in the junction area.

Based on these experiments, we proposed a local leakage model. Area dependence data having local leakage current can be explained with this model. As shown in Fig. 4, large, medium, and small-sized junctions were fabricated on a Si wafer, and many localized leakage spots were assumed to exist at random points on the wafer. Large junctions always have many local leakage spots. However, some of the medium and small-sized junctions have no local leakage spots. About 25% of the medium-sized junctions and 88% of the small junctions have no local leakage. These results are quite

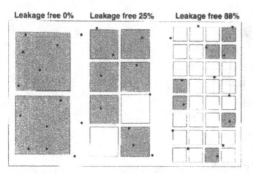

Leakage free 0%   Leakage free 25%   Leakage free 88%

Fig. 4. General description of the local leakage model.

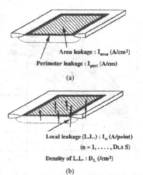

Area leakage : I_{area} (A/cm²)

Perimeter leakage : I_{peri} (A/cm)

(a)

Local leakage (L.L.) : I_n (A/point)

(n = 1, ... , D_L x S)

Density of L.L. : D_L (/cm²)

(b)

Fig. 5. Illustration of leakage current model (a) before and (b) after Co salicide process.

similar to the experimental data of the area dependence (Fig. 3).

Simulated Results and Localized Leakage Model

In order to verify our local leakage model, we simulated the leakage current using the Monte Carlo method. Fig. 5(a) and (b) shows illustrations of the leakage currents before and after Co salicide junctions. Generally, the leakage current is expressed as

$$I_{tot} = I_{area} \times S + I_{prei} \times L$$

where Iarea is the area leakage density and Iperi is the perimeter leakage density. In addition to this equation, we assumed the local leakage currents, therefore

$$I_{tot} = I_{area} \times S + I_{prei} \times L + I_1 + \cdots + I_n$$

where "n" equals with being the local leakage spot density on the Si wafer. In this simulation, we assumed that all local leakage currents are not equal, but may vary in strength. In our model, we used a Gaussian distribution. The local leakage currents were distributed with an average current of Iave (A/point) and a standard deviation of σ as shown in Fig. 6. The simulated results of the leakage current during the second RTA temperature of 750°C are shown in Fig. 7 and were extracted from a non-salicided junction. We only used three fitting parameters. The simulated results correspond very well with the experimental data to strongly support our local leakage model. In addition, this simulation also provided us with the characteristics of the local leakage current. In order to match the simulation data of three different areas with the experimental data, the three fitting parameters were uniquely determined. At 750°C, and were 4000 points/cm2 , 1E-12 A/point, and two orders, respectively. As we

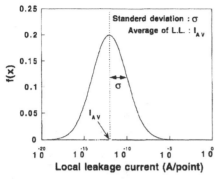

Fig. 6. Gaussian distribution of local leakage currents.

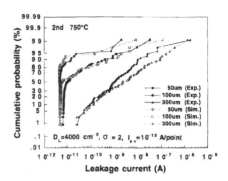

Fig. 7. Simulated results of leakage currents at second RTA temperature of 750 °C.

will discuss later, it is interesting that and always resulted in the same value, despite the difference in RTA temperatures from 400°C to 850°C. Only the density showed a change. As we have mentioned, it is clear that the leakage current induced by the Co salicide is generated from many localized spots.

Origin of the Localized Leakage Current

In the above discussion, we described the leakage current characteristics of the Co salicided junction. In this chapter we will discuss the origin of the leakage current, that is, when and why localized leakage currents are induced. We investigated the junction leakage current by varying the first RTA temperature to determine when the local leakage occurs. Fig. 8 shows the first RTA dependence of the leakage current at from 400°C to 550°C for 30 s. The leakage current changed drastically when varying the first RTA. It increased rapidly at between 400°C and 450°C. An annealing temperature of 450°C produced the maximum leakage current. Beyond 500°C, these leakage currents gradually reduced as the annealing temperature was increased. Fig. 9 shows an X-ray analysis of the Co silicide films under varying annealing conditions. 450°C was the temperature to form the Co Si phase from Co film. Accordingly, it is thought that local leakage is generated when Co2Si is formed. We also simulated all leakage currents during the first RTA as shown in Fig. 10, and a very good agreement was obtained. The densities were calculated to be 200, 100 000, and 6000 (points/cm2 for 400, 450, and 550°C, respectively, and the average and the standard deviation were always the same value at 10 A/point and two orders, respectively.

Fig. 11 shows the cross-sectional TEM images of Co salicided junctions at annealing temperatures of 400, 450, 550, and 825°C for 30 s. In the junction annealed at 400°C, several small spikes were observed under the Co silicide film. In the junction at 450°C, these spikes became larger at between 20 and 100 nm in length, which is enough to break shallow junctions. We observed the Co signal from these spikes by EDX analysis (Fig. 12), therefore we assumed these spikes to be a kind of

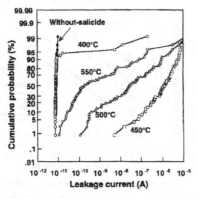

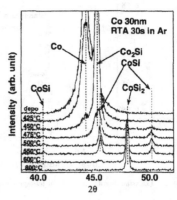

Fig. 8. First RTA dependence of the leakage current from 400 to 550 °C
for 30 s.

Fig. 9. X-ray diffraction of Co silicide films.

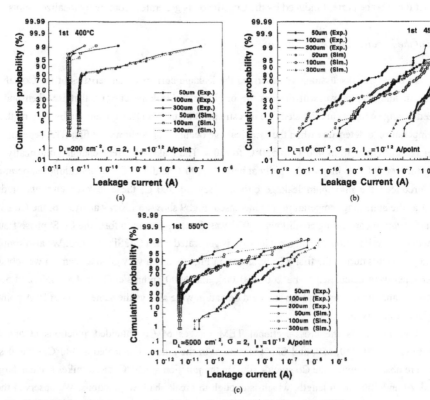

Fig. 10. Simulated results of leakage currents at first RTA temperature of (a) 400, (b) 450, and (c) 550 °C.

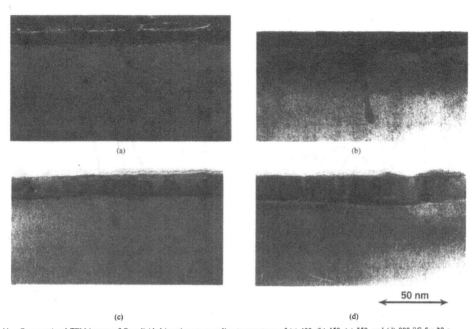

(a)   (b)

(c)   (d)

50 nm

Fig. 11.   Cross-sectional TEM images of Co salicided junctions at annealing temperatures of (a) 400, (b) 450, (c) 550, and (d) 800 °C for 30 s.

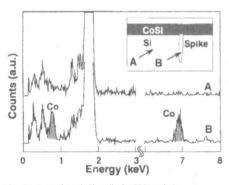

Fig. 12.   Co signals from CoSi$_x$ spike by EDX analysis.

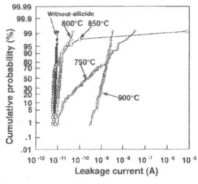

Fig. 13.   Leakage current with various second annealing temperatures at between 750 and 900 °C for 30 s (first RTA is fixed to 550 °C).

Co silicide. After annealing at 550°C, these spikes became smaller and more rounded. They disappeared altogether after annealing at over 800°C. The behavior of the CoSix spikes at various temperatures is very similar to that of the leakage current. Therefore, we consider these CoSix spikes to be the origin of the local leakage currents.

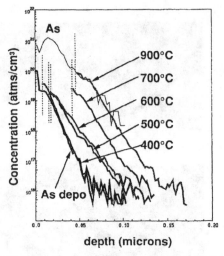

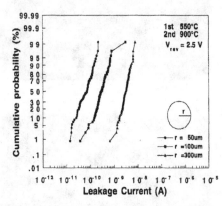

Fig. 14. SIMS profile of the Co atoms which diffused from Co silicide film during annealing at between 400 and 900 °C for 30 s.

Fig. 15. Area dependence of leakage current after second annealing at 900 °C.

Optimized Salicide Process Conditions

The leakage current at various second annealing temperatures of between 750°C and 900°C for 30 s were then investigated, as shown in Fig. 13. Conventional second annealing temperature at 750°C sufficiently reduces the sheet resistance but does not minimize the leakage current. Annealing a between 800°C and 850°C drastically reduces the leakage current down to almost the same level of a junction without salicide. However, further annealing at 900°C increased the leakage current again. This is because at such a high temperature, the CoSi2 film begins to melt and many Co atoms are subsequently diffused into the Si substrate.

To investigate the Co diffusion phenomenon, we carefully analyzed Co diffusion profiles annealed at between 400°C and 900°C via SIMS (Fig. 14). In order to prevent the knocking Co atoms in the Si substrate during the SIMS analysis, samples were analyzed after completely removing the Co silicide film by an HF treatment. The profile of arsenic atom, which was implanted to form a pn-junction, was used as a marker in order to compare them at the same original depth. Although Co diffusion was observed between 700°C and 800°C, it did not cause the junction leakage, because few Co atoms diffused compared to the arsenic atoms. At 900°C annealing, however, the Co atoms increased to comparable levels with the arsenic atoms, and such amounts of Co may have degraded the junction leakage current. At over 900°C annealing, a different leakage mechanism exists, as we explained. Fig. 15 shows the area dependence of the leakage current after the second annealing at 900°C. The leakage current distribution for all junction areas was small and did not depend on the area

This indicates that this leakage current was uniformly generated over the whole junction area. Therefore the leakage mechanism is different between below 850°C and over 900°C. The former mechanism is the localized leakage arising from the CoSix spikes, and the latter is the uniform area leakage due to the Co diffusion. By considering these two leakage mechanisms, the optimal conditions for the second annealing to minimize the leakage current exists between 800°C and 850°C, which minimizes the density and length of the CoSix spikes without causing Co diffusion. Using this optimal second RTA condition of 850°C, good characteristics were obtained, as shown in Fig. 16.

To consider the applicability of this optimized process to deep sub-micron CMOS devices, we estimated the characteristics of the local leakage currents via Monte Carlo simulation. Fig. 17 shows the simulation results for temperatures between 400°C and 900°C. In all temperatures except 900°C, the average current for the local leakage and the standard deviations are always the same at 1E-12 A/point and two orders, respectively.

Only the local leakage spot density changed. The density sharply increased at 450°C, and gradually decreased at annealing over 500°C. Employing the optimized second RTA condition of between 800°C and 850°C, the density of the local leakage spots decreased by only 50 to 100/cm2 Considering that a single local leakage is a very low current, we can conclude that the optimized Co salicide process never affects the operation of sub-0.1-um CMOS devices.

CONCLUSION

We clarified a new leakage mechanism of Co salicided shallow p-n junctions. Electrical measurements of the junction area dependence indicated that the leakage current flows from many localized points. We successfully simulated this leakage current with the local leakage model. We found CoSi spikes of abnormal growth under the Co silicide film to be the origin of the localized leakage currents. These CoSi spikes grew rapidly at annealing temperatures between 400°C and 450°C for 30 sec when Co2Si was formed, and drastically fell at annealing between 800°C and 850°C for 30 sec. Under this optimized second annealing condition, the density of local leakage points and the average of local leakage currents were calculated to be 50–100/cm2 and 10E-10 A/point, respectively. Because a single local leakage current is very low, this optimized Co salicide process never affects the operation of sub-0.1-um CMOS devices.

ACKNOWLEDGMENT

The authors wish to thank A. Fushida, J. Watanabe, T. Sukegawa, T. Yamazaki, and T. Sugii for their useful discussion and a lot of experiments.

REFERENCES

[1] T. Yamazaki, K. Goto, T. Fukano, Y. Nara, T. Sugii, and T. Ito, "21 ps switching 0.1-um-CMOS at room temperature using high performance Co salicide process," in IEDM Tech. Dig., 1993, pp 906–909.

[2] K. Goto, T. Yamazaki, A. Fushida, S. Inagaki, and H. Yagi, "Optimization of salicide process for 0.1-um CMOS devices," in Symp. VLSI Tech., 1994, pp. 119–120.

[3] J. B. Lasky, J. S. Nakos, O. J. Cain, and P. J. Geiss, "Comparison of transformation to low-resistivity phase and agglomeration of TiSi2 and CoSi2 ;" IEEE Trans. Electron Devices, vol. 38, pp. 262–271, Feb. 1991.

[4] A. C. Berti and V. Bolkhovsky, "A manufacturable process for the formation of self aligned cobalt silicide in submicronmeter CMOS technology," VMIC Conf. Dig., 1992, pp. 267–273.

[5] K. Goto, A. Fushida, J. Watanabe, T. Sukegawa, K. Kawamura, T. Yamazaki, and T. Sugii, "Leakage mechanism and optimized conditions of Co salicide process for deep submicron CMOS devices," in IEDM Tech. Dig., 1995, pp. 906–909.

Mat. Res. Soc. Symp. Vol. 611 © 2000 Materials Research Society

Modeling of Self-Aligned Silicidation in 2D and 3D:
Growth Suppression by Oxygen Diffusion

Victor Moroz and Takako Okada[*]
Avant! Corporation, 46871 Bayside Parkway, Fremont, CA 94538, U.S.A.
[*] Toshiba Corporation, Research and Development Center,
1 Komukai-Toshiba-cho, Kawasaki 210, Japan.

ABSTRACT

Stress-strain effects and physical processes during formation of the self-aligned silicides are analyzed. A new model for predictive simulation of the self-aligned silicidation is suggested. The model is based on suppression of diffusion and reaction rate of the silicon atoms inside silicide in the presence of oxygen atoms, injected into silicide from the neighbor oxide regions such as oxide spacer, TEOS at STI (Shallow Trench Isolation) and pad oxide. The model is demonstrated to explain the experimentally observed silicide shape.

INTRODUCTION

Self-aligned silicidation is a commonly accepted method of reducing series resistance of the source, drain, and gate terminals of deep submicron MOSFETs. Typical experimentally observed $TiSi_2$ shape near an oxide spacer is shown in Figure 1.

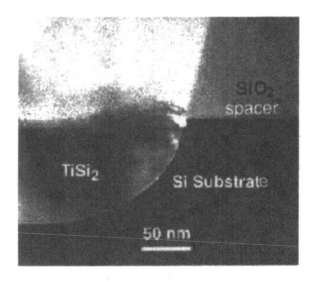

Figure 1. TEM image of self-aligned titanium silicide next to oxide spacer from [1]. Silicidation was performed at $730^{\circ}C$.

There is virtually 100% suppression of the lateral silicidation under the spacer. This observation did not receive a satisfactory theoretical explanation, yet its understanding is necessary for predictive simulation of the silicide shape, which determines source/drain impurity diffusion and stress distribution around the silicide layer.

Simulated TiSi$_2$ shape is shown in Figure 2 using the same process flow as in the experiment, depicted in Figure 1. Lateral silicidation here is comparable to the vertical silicidation due to the assumption of constant diffusivity for the silicon atoms, which are the dominant diffusing species in TiSi$_2$. Similar shape is obtained also if metal atoms are the dominant diffusing species, or if several species are simultaneously contributing to the silicide growth, as long as their diffusivities and reaction rates are constant throughout the silicide.

THEORY

It has been suggested that mechanical stress can be responsible for suppressing the silicide growth at the spacer corner [1]. However, stress-strain analysis of silicidation shows that stress changes sign near the end of the spacer (see, for example Figure 2). Such a stress pattern in silicide is due to the downward movement of the shrinking metal layer next to the spacer. The entire metal layer is moving down to provide metal atoms for the growing silicide. The movement is affected by the adjacent oxide spacer, which slows down the metal flow along the spacer, but in turn pushes down the silicon substrate below.

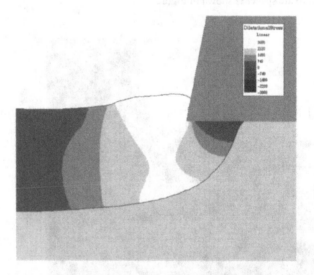

Figure 2. Simulated self-aligned titanium silicidation at 730°C. Hydrostatic pressure is shown in MPa at the silicidation temperature. Silicon atoms diffusivity and reaction rate are constant.

Similar stress distribution has been obtained in [1]. Such a stress pattern with stresses of the opposite sign is inherent for the self-aligned silicidation and does not depend on particular metal or the dominant diffusing species, but rather is determined by geometry and physics common to all silicides.

If we introduce stress dependent diffusivity of the diffusing species in a layer with stresses of the opposite signs, then the diffusivity will be inevitably suppressed in one part of the layer and enhanced in the other, whereas measurements show monotonic change of the growth rate vs. distance to the spacer.

It is known that implanted oxygen [2], oxygen, incorporated into metal during deposition [3], or oxygen from TiO_x/SiO_2 [4] suppresses growth of different types of silicides, including $TiSi_2$. We suggest that growth rate of the self-aligned silicide near the oxide spacer is reduced due to injection of oxygen atoms from SiO_2 into the silicide and suppression of silicon atom diffusivity and reaction rate. Oxygen diffusion in silicide is simulated as a combination of the bulk and boundary diffusion using a generalization of the interface trapping model [5].

RESULTS

Simulated $TiSi_2$ shape using the suggested model is shown in Figure 3.

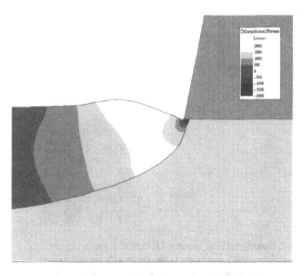

Figure 3. Simulated self-aligned titanium silicidation at 730°C. Hydrostatic pressure is shown in MPa at the silicidation temperature. Silicon atoms diffusivity and reaction rate depend on the local oxygen concentration.

There is no lateral silicidation under the spacer just like in the measurements. Simulation of the self-aligned silicidation with nitride spacer gives similar silicide shape due to the oxygen injection from the thin pad oxide under the nitride spacer.

Stress distribution in silicon substrate under the silicide in between the oxide spacer and STI (Shallow Trench Isolation) is shown in Figure 4 (2D) and in Figure 5 (3D) after cooling down the wafer and removing residual metal. Stresses generated during silicidation, and especially due to the thermal mismatch provide a major contribution to the total stresses in the MOSFET's channel next to the source and drain p/n junctions. These stresses are known to increase junction leakage current through the band gap narrowing.

CONCLUSIONS

A new model for predictive simulation of the self-aligned silicidation has been suggested. The model adequately describes shape of the self-aligned silicides, as verified by comparison to TEM images. 2D and 3D simulations have been performed to investigate stress-strain evolution during the process flow and its impact on device performance.

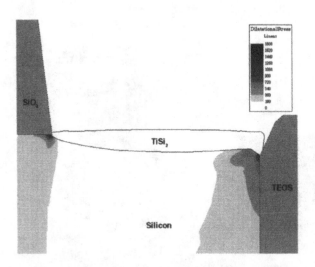

Figure 4. Simulated TiSi$_2$ over a MOSFET's drain region. Residual hydrostatic pressure is shown in silicon substrate in MPa after cooling the wafer down to room temperature and removing residual metal.

REFERENCES

1. P. Fornara and A. Poncet, *International Electron Devices Meeting Technical Digest*, 73-76 (1996).
2. F. Nava S. Valeri, G. Majni, A. Cembali, G. Pignatel and G. Queirolo, *J. Appl. Phys.*, **52**, 6641-6646 (1981).
3. H. Jiang, H. J. Whitlow, M. Ostling, E. Niemi, F. M. d'Heurle and C. S. Petersson, *J. Appl. Phys.*, **65**, 567-574 (1989).
4. S. W. Russell, J. W. Strane, J. W. Mayer and S. Q. Wang, *J. Appl. Phys.*, **76**, 257-263 (1994).
5. Y.-S. Oh and D. E. Ward, *International Electron Devices Meeting Technical Digest*, 509-512 (1998).

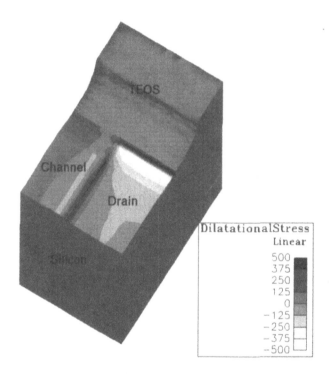

Figure 5. *Simulated TiSi$_2$ over a MOSFET's drain region. Residual hydrostatic pressure is shown in silicon substrate in MPa after cooling the wafer down to room temperature and removing residual metal. The gate stack and silicide layer are not shown to see stresses in the silicon substrate. The 3D corner of the silicide apparently has lower stress values than the 2D edges of the silicide along the oxide spacer and STI.*

Epitaxial Silicides

Mat. Res. Soc. Symp. Vol. 611 © 2000 Materials Research Society

Epitaxial CoSi$_2$ formation by a Cr or Mo interlayer.

C. Detavernier, R.L. Van Meirhaeghe, F. Cardon
Vakgroep Vaste Stofwetenschappen, Universiteit Gent, Krijgslaan 281/S1, B-9000 Gent, Belgium.

K. Maex[+], B. Brijs, W. Vandervorst
IMEC, Kapeldreef 75, B-3001 Leuven, Belgium.
[+] also at E.E. Dept, K.U. Leuven, B-3001 Leuven, Belgium.

ABSTRACT

We have studied CoSi$_2$ formation in the presence of a Cr or Mo interlayer or capping layer. We shall show that, contrary to what was previously reported, Cr and Mo may be used as interlayers to grow epitaxial CoSi$_2$. However, unlike for Ti, the thickness of the interlayer is very important. If the Cr or Mo interlayer is too thick (> 5 nm), polycrystalline CrSi$_2$ or MoSi$_2$ are formed first and epitaxial growth of CoSi$_2$ becomes impossible. However, both XRD and random/channeling RBS results indicate that for a 2-3 nm interlayer of Cr or Mo, CoSi$_2$ forms epitaxially on Si(100). For thinner interlayers, there is a preferential (220) and (400) orientation. This can be explained by the presence of Cr or Mo on the CoSi grain boundaries, which will affect the heterogeneous nucleation of CoSi$_2$.

INTRODUCTION

Due to the small lattice mismatch (-1.2%), CoSi$_2$ may be formed epitaxially on Si. However, in the standard Co/Si reaction, polycrystalline CoSi$_2$ is formed. Various methods of obtaining epitaxial CoSi$_2$ growth have been developed over the last decades : techniques based on molecular beam epitaxy (template methods [1], allotaxy [2]), ion beam synthesis (IBS) by high dose Co implantation [3] and solid phase epitaxy (SPE), using an interlayer in between the Co and Si substrate [4].

The presence of a Ti interlayer is known to result in epitaxial growth of CoSi$_2$. In literature, the epitaxial growth is usually explained by (1) the ability of Ti to reduce interfacial SiO$_2$ and by the fact that (2) the Ti interlayer acts as a diffusion barrier, mediating the Co flux towards the substrate. Inspired by the results for the Co/Ti/Si system and in an attempt to determine the mechanism responsible for solid phase epitaxy (SPE), several groups have studied the effect of other interlayer materials. It has been shown that SPE "works" for Ta, Hf, C, W and Zr interlayers, while in case of Mo [5], V and Cr [6,7] interlayers, it was reported that the solid phase reaction results in polycrystalline CoSi$_2$.

In this paper, it is shown that the previous results on Cr and Mo interlayers should not be interpreted as "Cr and Mo are bad interlayer materials to promote epitaxy" : it is shown that depending on the thickness of the interlayer, either polycrystalline or epitaxial CoSi$_2$ can be formed. It will be shown that the amorphous character of the diffusion mediating interlayer is important to achieve epitaxial growth of the CoSi$_2$.

EXPERIMENTAL

We used p-type Si(100) substrates ($N_a = 10^{13}$-10^{14} cm^{-3}). Cr and Co were deposited by e-beam evaporation in a vacuum of 10^{-6} mbar. Mo was deposited by DC sputtering in argon. Deposition of the multilayer structure was done without breaking the vacuum. After deposition, the wafers were cut into smaller pieces and annealed in a Rapid Thermal Processing (RTP) system in N_2 ambient. To study the silicidation reaction, isochronal annealing (30 seconds) was done at various temperatures.

After annealing, sheet resistance measurements and grazing incidence XRD were used for phase identification. XPS depth profiling was applied to study the chemical composition of the layers. Standard θ/2θ XRD and RBS (Rutherford Backscattering Spectroscopy) were used to measure preferential crystallographic orientation and epitaxial growth.

RESULTS AND DISCUSSION

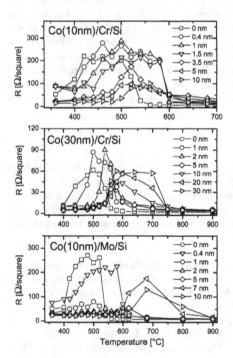

Figure 1 : Sheet resistance versus annealing temperature for Co/Cr/Si and Co/Mo/Si structures.

Phase formation

Phase formation was studied by sheet resistance and grazing incidence XRD. In figure 1, the sheet resistance is plotted versus the annealing temperature for the Co/Cr/Si and Co/Mo/Si system. For a thin Cr interlayer, CoSi forms at about 400°C, as evidenced by the sheet resistance and XRD (figure 1a and 2). At higher temperature (about 600°C), the CoSi is transformed into $CoSi_2$. For thicker Cr interlayers (> 4 nm), Co cannot penetrate the interlayer and the bottom part of the interlayer reacts with the Si substrate to form $CrSi_2$ (at about 500°C). Once the $CrSi_2$ is formed, CoSi is formed on top of it, as evidenced by XRD and XPS (figure 3). At higher temperature (> 600°C), the CoSi is transformed into $CoSi_2$, while the $CrSi_2$ remains present.

The fact that this $CrSi_2$ layer acts as a diffusion mediating layer is illustrated by figure 1b : the thicker the Cr interlayer, the higher the temperature needed to form the low-resistive $CoSi_2$ layer. This may be explained by the following argument : for a thicker Cr interlayer, more $CrSi_2$ is formed. In order to form $CoSi_2$, diffusion is needed through the $CrSi_2$ layer, thus the thicker the interlayer, the slower the $CoSi_2$ layer will be able to grow.

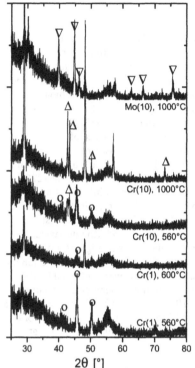

Figure 2 : Grazing incidence XRD for Co(10nm)/Cr/Si and Co(10nm)/Mo/Si structures after annealing : CoSi(o), CoSi₂ (□), CrSi₂(Δ) and MoSi₂(∇).

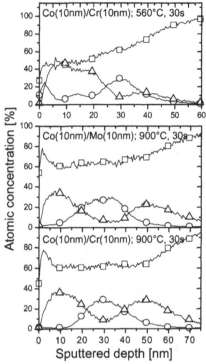

Figure 3 : XPS depth profiles for a Co/Mo and Co/Cr bilayer annealed at 560 and 900°C. The atomic concentration of Si(□), Co(Δ) and Mo or Cr (o) is plotted.

In the case of a Mo interlayer, similar behavior is observed as for Cr (figure 1c, 2). It was found that for $d_{Mo} < 2nm$, CoSi is formed between 400-500°C (depending on the interlayer thickness), followed by CoSi₂ formation at about 600°C. For $d_{Mo} > 2nm$, Co cannot diffuse through the interlayer sufficiently and the bottom part of the Mo interlayer reacts with the Si substrate at about 600°C to form MoSi₂, after which CoSi is formed on top of the MoSi₂. Only after annealing at 800°C, the CoSi is transformed into CoSi₂.

XPS was used to determine the atomic concentration as a function of depth (figure 3). For a 10 nm Cr interlayer, annealing results in a CoSi₂/CrSi₂/CoSi₂/Si structure. For Mo, a similar behavior was observed : for a Co(10)/Mo(10)/Si structure, annealing at 900°C resulted in a CoSi₂/MoSi₂/CoSi₂/Si structure.

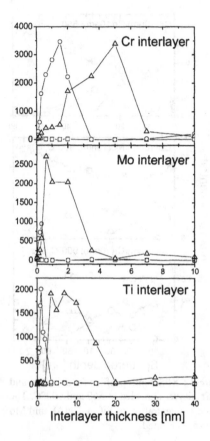

We used θ/2θ XRD to measure the preferential orientation of the CoSi₂. From figure 4, it is clear that the preferential orientation is dependent on the thickness of the interlayer : for the reference sample (standard Co/Si reaction without any interlayer), a polycrystalline CoSi₂ layer is formed, for which (111) and (220) peaks are visible in XRD. In the presence of a thin interlayer (< 2nm), we observed a strong increase of the (220) and (400) peak and a decrease of the (111) peak. For thicker interlayers, the (220) peak intensity decreases, while the (400) intensity increases further, indicating epitaxial alignment of the CoSi₂ with the Si substrate. However, for thick Cr and Mo interlayers, XRD indicates that the epitaxial alignment disappears. The variation of epitaxial alignment as a function of interlayer thickness was confirmed by RBS, both for Cr (minimum channeling yield of 42% for a Co(10nm)/Cr(3.5nm)/Si bilayer annealed at 900°C) and Mo (minimum yield of 39% for a Co(10nm)/Mo(2nm)/Si bilayer annealed at 900°C).

For comparison, we also studied the preferential orientation for a Co(50nm)/Ti/Si system. The Co layer thickness was selected to allow the use of thicker Ti interlayers than normally used for TIME. It is found that qualitatively, the same type of behavior is found for Ti interlayers as was observed for Cr and Mo interlayers : there are three regimes with a different preferential orientation, depending on the thickness of the interlayer.

Figure 4 : Intensity of the CoSi₂ (111) (□), (220) (o) and (400) (Δ) XRD peak as a function of interlayer thickness, for a Cr, Mo and Ti interlayer.

CONCLUSION

It has been reported previously for Ti [8] that there is a certain critical thickness for the interlayer which should not be exceeded in order to avoid $TiSi_2$ formation. In this work, it is found that because of the lower formation temperature of $CrSi_2$ and $MoSi_2$, the critical thickness is much smaller for these materials. Thus, the succes of using Cr or Mo for SPE will be much more strongly dependent on the interlayer thickness than for e.g. Ta, Zr, W and Ti. However, except for the smaller value for the critical thickness, there does not seem to be a fundamental difference in the reaction sequence and in the behavior of preferential orientation versus interlayer thickness for Cr and Mo as compared to other interlayers, such as Ti. This is an important indication that can help to shed some light upon the mechanism underlying SPE : first of all, one may conclude from this that the ability of the metal to reduce SiO_2 is probably not as important as was believed until now. Secondly, it seems that the interlayer material may directly influence the $CoSi_2$ nucleation. This effect is observed most clearly for very thin interlayers, while for thicker interlayers it is probably still present, but it is no longer distinguishable from the diffusion mediating effect of the interlayer.

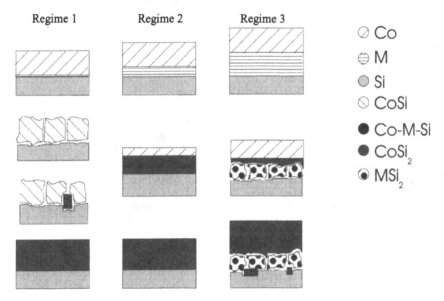

Figure 5 : Schematics of reaction model for $CoSi_2$ formation in the presence of an interlayer material M, illustrating the three different reaction regimes.

It seems that SPE works because of two different mechanisms that play in two different regions of parameter space : for thin interlayers, the interlayer should not be thought of as a diffusion mediating layer, but mainly as a 'contaminant' present in the CoSi phase. It has been shown that e.g. Ti is able to delay nucleation and to cause preferential orientation. This was explained by the ability of Ti to improve the cohesion of the CoSi grain boundaries[9]. For thicker interlayers, a second mechanism becomes important : the interlayer mediates the diffusion of the Co towards the Si substrate. For this second regime, it is important that the interlayer metal reacts with the Co and Si to form preferably an amorphous interlayer, which is able to slow down Co diffusion in a laterally homogeneous way. Once the interlayer thickness has exceeded a certain critical value (third regime), the interlayer reacts with the Si to form a polycrystalline disilicide. For this regime, it has been observed in this work that polycrystalline $CoSi_2$ is formed on top of and underneath the disilicide interlayer. The formation of $CoSi_2$ on top of the MSi_2 can be explained based on the fact that diffusion of Si in disilicides is generally faster than diffusion of metals. The polycrystalline nature of the disilicide interlayer allows for fast grain boundary diffusion through the interlayer, resulting in the formation of some $CoSi_2$ underneath the disilicide interlayer. The bottom $CoSi_2$ layer is polycrystalline, in spite of the MSi_2 diffusion barrier, since the grain boundaries allow faster Co diffusion towards the substrate than would be the case for an amorphous interlayer.

ACKNOWLEDGEMENTS

The authors would like to thank Ing. L. Van Meirhaeghe for technical support and Ing. U. Demeter for XPS depth profiling measurements. C. Detavernier thanks the 'Fonds voor Wetenschappelijk Onderzoek – Vlaanderen' (FWO) for a scholarship. K. Maex is a research director for the FWO.

REFERENCES

1. R.T. Tung, F. Schrey, MRS Symp. Proc. **122**, 559 (1988).
2. S. Mantl, H.L. Bay, Appl. Phys. Lett. **61**, 267 (1992).
3. A.E. White, K.T. Short, R.C. Dynes, J.P. Garno, J.M. Gibson, Appl. Phys. Lett. **50**, 95 (1987).
4. M. Lawrence, A. Dass, D.B. Fraser, and C.S. Wei, Appl. Phys. Lett. **58**, 1308 (1991).
5. S.-L. Zhang, F.M. d'Heurle, summerschool on silicides, Erice (1999).
6. J.S. Byun, H.J. Kim, J. Appl. Phys. **78**, 6784 (1995).
7. J.S. Byun, W.S. Kim, M.S. Choi, H.J. Cho, H.J. Kim, MRS Proc. **320**, 379 (1994).
8. F. Hong, G.A. Rozgonyi, J. Electrochem. Soc. **141**, 3480 (1994).
9. C. Detavernier, R.L. Van Meirhaeghe, F. Cardon, K. Maex, MRS Proc. Spring Meeting 2000 (also in this volume).

Mat. Res. Soc. Symp. Vol. 611 © 2000 Materials Research Society

IN-SITU GROWTH AND GROWTH KINETICS OF EPITAXIAL (100) $CoSi_2$ LAYER ON (100) Si BY REACTIVE CHEMICAL VAPOR DEPOSITION

Hwa Sung Rhee, Heui Seung Lee, Jong Ho Park, and Byung Tae Ahn
Department of Materials Science and Engineering, Korea Advanced Institute of Science and Technology, 373-1 Koosung-dong, Yusung-gu, Taejon 305-701, Korea

ABSTRACT

Uniform epitaxial $CoSi_2$ layers have been grown *in situ* on a (100) Si substrate at temperatures near 600 °C by reactive chemical-vapor deposition of cyclopentadienyl dicarbonyl cobalt, $Co(\eta^5\text{-}C_5H_5)(CO)_2$. The growth kinetics of an epitaxial $CoSi_2$ layer on a Si (100) substrate was investigated at temperatures ranging from 575 to 650 °C. In initial deposition stage, plate-like discrete $CoSi_2$ spikes were nucleated along the <111> directions in (100) Si substrate with a twinned structure. The discrete $CoSi_2$ plates with both {111} and (100) planes grew into an epitaxial layer with a flat interface on (100) Si. For epitaxial $CoSi_2$ growth on (100) Si, the activation energy of the parabolic growth was found to be 2.80 eV. The growth rate seems to be controlled by the diffusion of Co through the $CoSi_2$ layer.

INTRODUCTION

As the microelectronics device dimensions scale down to the deep submicron level, cobalt disilicide ($CoSi_2$) is the most promising due to its low resistivity and linewidth-independent sheet resistance. Epitaxially grown layers of $CoSi_2$ on Si (100) substrate are of special interest because of their excellent thermal stability, low junction leakage, and ultra-shallow junction formation using the silicide-as-doping-source process[1]. For their possible applications in microelectronic devices such as ohmic contacts, low resistivity gates, and local buried interconnections, the growth kinetics of epitaxial $CoSi_2$ on Si (100) substrate need to be investigated. Due to its small lattice mismatch with respect to Si (-1.2%) and cubic CaF_2 structure, $CoSi_2$ can be epitaxially grown on Si substrates provided the film thickness does not exceed the critical thickness. Despite the promising structural match of $CoSi_2$ and Si, the growth of an epitaxial $CoSi_2$ layer on (100) Si substrate has not been successfully realized due to the strong tendency of growth of misoriented grains[1]. Therefore, the special growth techniques such as interlayer mediated epitaxy, molecular beam epitaxy, ion beam synthesis, and reactive-deposition epitaxy have to be applied to produce $CoSi_2$/Si heteroepitaxy structure[2-4].

Recently, we reported the method of *in situ* growth of epitaxial $CoSi_2$ layer on (100) Si substrate by reactive chemical-vapor deposition using a cobalt metallorganic source of cyclopentadienyl dicarbonyl cobalt, $Co(\eta^5\text{-}C_5H_5)(CO)_2$[5]. In this technique, a uniform epitaxial $CoSi_2$ layer with good quality can be *in situ* grown on (100) Si substrate heated above 600 °C by controlling the Co flux in a simple metallorganic chemical vapor deposition reactor without multi-step deposition and subsequent annealing. In this study, the growth behavior and kinetics of the epitaxial $CoSi_2$ layers grown on Si (100) substrate by reactive chemical vapor-deposition were investigated at temperatures ranging 575 to 650 °C.

EXPERIMENTAL

The p-type (100) oriented Si substrates with a resistivity of 5-8 Ω·cm were cleaned in a H_2SO_4/H_2O_2 solution, rinsed in deionized water, dipped in HF (1%), rinsed again in deionized water, and then loaded into a MOCVD reactor. The temperature of the bubbler was held at - 5 °C to reduce the vapor pressure, allowing the supply of Co into the substrate to be held constant to form the epitaxial $CoSi_2$. The Co was supplied to the Si substrate heated to temperatures ranging from 575 to 650 °C using $Co(\eta^5-C_5H_5)(CO)_2$ at 110 mTorr with 10 sccm H_2 carrier gas. The microstructure of the films was observed using transmission electron microscopy in plan-view and cross-section mode. The thicknesses of the epitaxial $CoSi_2$ layers were calculated from the measured sheet resistance values. The electrical sheet resistances of the films were monitored by a four-point probe method.

RESULTS AND DISCUSSION

Figure 1 shows the XTEM micrograph of the sample as-grown at 650 °C for 30 min by reactive chemical vapor deposition epitaxy. An amorphous Si layers were deposited at room temperature by PECVD with a argon diluted SiH_4 gas after the epitaxial growth of $CoSi_2$ in order to measure the thickness of the thin carbon layer. Since the epoxy layer to prepare the XTEM specimen and the carbon layer are all the amorphous carbon, the difference in contrast between the layers almost does not exist in XTEM micrograph. The 6.2 nm thick carbon layer is about 1/10 of the epitaxial $CoSi_2$ layer.

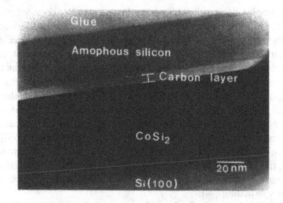

Fig. 1. XTEM micrograph of the sample as-grown at 650 °C for 30 min by RCVDE with an amorphous Si capping layer.

Figure 2 shows the electrical sheet resistances of the films measured after deposition at 650 °C for different times. The resistivity ρ and the sheet resistance Rs of a thin film are related by Rs = ρ/d, where d is the thickness of the thin film. $CoSi_2$ has the lowest resistivity (~10 $\mu\Omega$·cm for the epitaxial layer and 15-18 $\mu\Omega$·cm for polycrystalline film) among all the silicides. However, Hirano et al. and Hensel et al. reported that the resistivity of the single-crystal $CoSi_2$ were measured 17.2 and 15 $\mu\Omega$·cm at room temperature due to high residual resistivity, respectively[6, 7]. The high residual resistivity is attributed to the existence of intrinsic disorder related to defects originating in the formation of epitaxial $CoSi_2$ layer. The thicknesses of the $CoSi_2$ layers were measured from the cross-sectional TEM micrographs of the samples (15, 30, and 40 min). Then, the resistivity of the epitaxial $CoSi_2$ layer was calibrated 17.3 ± 0.1 $\mu\Omega$·cm from the sheet resistance and the layer thickness, in good agreement with Hirano et al.'s result.

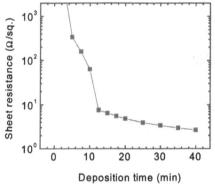

Fig. 2. Electrical sheet resistance of the $CoSi_2$ films measured after deposition at 650 °C for different times.

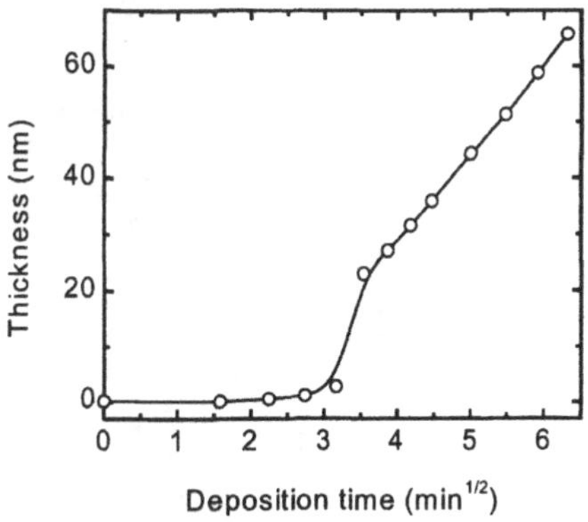

Fig. 3. Thickness of the epitaxial $CoSi_2$ layers calculated from the sheet resistance of the films as a function of deposition time.

Figure 3 shows the thickness of the epitaxial $CoSi_2$ layers calculated by the sheet resistances of the films as a function of the deposition time, postulating that the epitaxial $CoSi_2$ layers have a uniform thickness and the same resistivity of 17.3 $\mu\Omega$·cm. The calculated data are plotted versus the square root of the deposition times, where the two regions exist in Fig. 3. In the initial incubation or nucleation period, the $CoSi_2$ thickness does not increase rapidly with increasing deposition time. After the initial incubation period, the data well fitted by a straight line, indicating the parabolic growth relationship between the thickness of the epitaxial $CoSi_2$ and the deposition time.

To observe the growth behavior of epitaxial $CoSi_2$ on (100) Si substrate, $CoSi_2$ layers in the initial deposition stage were investigated using the plan-view and cross-sectional view TEM. Figure 4 (a) shows the plan-view TEM image of the sample grown at 650 °C for 5 min deposition along the <100> zone axis. Two-dimensional elongated $CoSi_2$ islands form and are aligned perpendicular to the <011> directions of the substrate. Their shape becomes more bar-like, or plate-like, accompanied by an increasing roughness of the free surface. The $CoSi_2$ nuclei grew to about 50~100 nm long rods with two different orientations. Scheuch et al. found by in situ scanning tunneling microscopy that a very low Co coverage first two-dimensionally elongated $CoSi_2$ islands forms aligned perpendicular to the <011> directions, in good agreement with the results of Fig. 4 (a)[8]. Figure 4 (b) shows the cross-sectional TEM image of the as-deposited sample at 650 °C for 5 min. The plate-like $CoSi_2$ spikes are randomly grown along the equivalent {111} interfaces on (100) Si substrate.

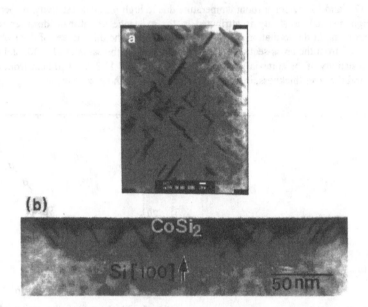

Fig. 4. (a) Plan-view TEM image and (b) cross-sectional TEM image of the sample grown at 650 °C for 5 min deposition.

The cross-sectional TEM micrographs of Figure 5 show that the epitaxial $CoSi_2$ layers after deposition at 650 °C for (a) 10, (b) 15, and (c) 40 min. The $CoSi_2$ spikes grow into a discrete $CoSi_2$ with {111} and (100)-faceted interfaces [Fig. 5(a)] and then the sheet resistance of the

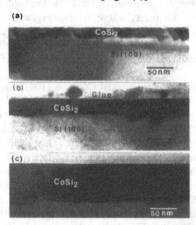

Fig. 5. XTEM micrographs of the samples after reactive deposition at 650 °C for (a) 10 min, (b) 15 min, and (c) 40 min.

C10.3.4

epitaxial CoSi$_2$ layer rapidly decreases due to the continuous layer formation. The uniform epitaxial layers with a flat (100) interface were formed on Si (100) substrate after prolonged deposition time [Fig. 5(b)], resulting in the parabolic growth[9]. The cross-sectional TEM of Fig. 5(c) shows the uniform epitaxial CoSi$_2$ layer with a thickness of about 70 nm with small {111} facets. The plate-like CoSi$_2$ nuclei in the initial stage cause a large interface roughness during growth and hinder the formation of very thin uniform layers of less than 10 nm thickness. In order to obtain smaller plate-like CoSi$_2$ nuclei, the substrate temperature must be lowered. Also a deposition rate increased enough to form epitaxial CoSi$_2$ layer produces smaller spikes due to an enhancement of the nucleation rate and suppression of the growth rate.

Figure 6 shows the epitaxial CoSi$_2$ thickness as a function of deposition time at temperatures ranging from 575 to 650 °C. To obtain the parabolic rate constants from these data, the basic kinetic equation and the scheme utilized by Deal and Grove were adopted. The equation is $x^2/k_P + x/k_L = t + \tau$, where x is the CoSi$_2$ thickness, t is the deposition time, k_P and k_L are the parabolic rate constant and linear rate constant, and τ is a time constant. The initial incubation period due to the growth of epitaxial CoSi$_2$ nuclei decreases rapidly with increasing temperature.

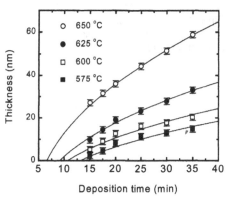

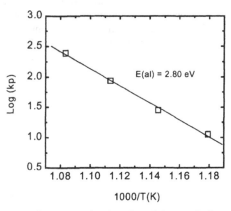

Fig. 6. Epitaxial CoSi$_2$ thickness as a function of deposition time at temperatures ranging from 575 to 650 °C.

Fig. 7. Arrehenius plot of the parabolic rate constants, k_P.

Figure 7 shows the Arrehenius plot of the parabolic rate constants, k_P, obtained from the well-fitted parameters. A good straight line fit is obtained with a slope that yields an activation energy of 2.80 eV. The activation energy of 2.80 eV for the growth of the epitaxial CoSi$_2$ layer almost corresponds to the reported value, 2.6-2.9 eV[10]. For CoSi$_2$, the diffusion coefficient of Co in the lattice has been reported as 0.15 exp (-2.78/kT)[11]. Therefore the activation energy of 2.80 eV indicates that the epitaxial growth of CoSi$_2$ is related to the diffusion-controlled growth.

CONCLUSION

The plate-like CoSi$_2$ spikes with a twinned structure were formed in the initial stage of reaction between Co and Si. The discrete CoSi$_2$ plates with {111} and (100) planes grew into a uniform epitaxial layer during reactive chemical-vapor deposition. In the initial incubation period, the CoSi$_2$ thickness did not rapidly increase with increasing deposition time. After the formation of a uniform CoSi$_2$ layer, the epitaxial CoSi$_2$ growth corresponded to the parabolic growth relationship. The epitaxial CoSi$_2$ growth is limited by the diffusion of Co through the CoSi$_2$ layer.

REFERENCES

1. K. Maex, Mater. Sci. Eng., **R. 11**, 53 (1993).
2. A. E. White, K. T. Short, R. C. Dynes, J. P. Garno, and J. M. Gibson, Appl. Phys. Lett. **50**, 95 (1987).
3. A. H. Reader, J. P. W. B. Duchateau, and J. E. Crombeen, Semicond. Sci. Techonol. **8**, 1204 (1993).
4. M. L. A. Dass, D. B. Fraser, and C. S. Wei, Appl. Phys. Lett. **58**, 1308 (1991).
5. H. S. Rhee and B. T. Ahn, Appl. Phys. Lett. **74**, 3176 (1999).
6. T. Hirano and M. Kaise, J. Appl. Phys. **68**, 627 (1990).
7. J. C. Hensel, R. T. Tung, J. M. Poate, and F. C. Unterwald, Appl. Phys. Lett. **44**, 913 (1984).
8. V. Scheuch, B. Boigtlander, and H. P. Bonzel, Surf. Sci. **372**, 71 (1996).
9. H. S. Rhee, B. T. Ahn, and D. K. Sohn, J. Appl. Phys. **86**, 3452 (1999).
10. A. Appelbaum, R. V. Knoell, and S. P. Murarka, J. Appl. Phys. **57**, 1880 (1985).
11. T. Barge, P. Gas, and F. M. d'Heurle, J. Mater. Res. **10**, 1134 (1995).

AUTHOR INDEX

SUBJECT INDEX

Printed in the United States
By Bookmasters